T0220483

Graphene and Carbon Nanotubes for Advanced Lithium Ion Batteries

Graphene and Carbon Nanotubes for Advanced Lithium Ion Batteries

By
Stelbin Peter Figerez
Raghavan Prasanth

CRC Press
Taylor & Francis Group
Boca Raton London New York

CRC Press is an imprint of the
Taylor & Francis Group, an **informa** business

CRC Press
Taylor & Francis Group
6000 Broken Sound Parkway NW, Suite 300
Boca Raton, FL 33487-2742

First issued in paperback 2021

ISBN-13: 978-1-138-35312-1 (hbk)
ISBN-13: 978-1-03-217847-9 (pbk)
DOI: 10.1201/9780429434389

Publisher's Note

The publisher has gone to great lengths to ensure the quality of this reprint but points out that some imperfections in the original copies may be apparent.

Library of Congress Cataloging-in-Publication Data

Names: Prasanth, Raghavan, author. | Figerez, Stelbin Peter, author.
Title: Graphene and carbon nanotubes for advanced lithium ion batteries /
Raghavan Prasanth, Stelbin Peter Figerez.
Description: Boca Raton : Taylor & Francis, a CRC title, part of the Taylor & Francis imprint, a member of the Taylor & Francis Group, the academic division of T&F Informa, plc, 2018. | Includes bibliographical references and index.
Identifiers: LCCN 2018037708| ISBN 9781138353121 (acid-free paper) | ISBN 9780429434389
Subjects: LCSH: Lithium ion batteries--Materials. | Graphene. | Carbon nanotubes.
Classification: LCC TK2945.L58 P73 2018 | DDC 621.31/2424--dc23
LC record available at https://lccn.loc.gov/2018037708

Visit the Taylor & Francis Web site at
http://www.taylorandfrancis.com

and the CRC Press Web site at
http://www.crcpress.com

Dedicated to our beloved Sindhu, Stedlin, Steleena, and the naughty little prince Aaron, without whom this book would have been completed in half of the time.

Contents

Foreword

MODERN SOCIETY RUNS ON the energy stored in a fossil fuel; this development is not sustainable. It is not possible to recycle fossil fuels once they are burned, and the exhaust gases of their combustion contribute to global warming and to serious air pollution in many large cities. Since the nuclear alternative has been found to be problematic, it has become apparent that society must return to dependence on the energy that comes to us from the sun. Since hydropower stored in dams is not sufficient and biomass is needed to feed life on Earth, society is forced to look to the energy stored in sunlight and wind; this energy can be converted and transported to distributed collection sites, but it cannot be used unless it can be stored. The grid can store only a fraction of this energy; the rechargeable battery is the best alternative. Carbons, particularly graphene and carbon nanotubes, have played an important role in the development of the lithium-ion battery and will continue to be used in tomorrow's alkali-metal batteries whether they have liquid or solid electrolytes. Graphene and carbon nanotubes combine multiple outstanding physical and chemical properties such as high electronic and thermal conductivities, excellent mechanical strength and flexibility, and a huge surface area. This book provides a history of the invention of the lithium-ion battery and brings together available information on the roles of graphene and carbon nanotubes in the development of advanced lithium-ion batteries.

Prof. John B Goodenough

Authors

**Dr. Raghavan Prasanth** is a professor in the Department of Polymer Science and Rubber Technology, Cochin University of Science and Technology (CUSAT). He received his PhD in Engineering from the Department of Chemical and Biological Engineering, Geyongsang National University, Republic of Korea, in 2009, under the prestigious Brain Korea (BK21) Fellowship. He completed his B.Tech and M.Tech at CUSAT, India. After a couple of years as a project scientist at the Indian Institute of Technology (IIT-D), New Delhi, he moved abroad to complete his PhD studies in 2007. His PhD research was focused on the fabrication and investigation of nanoscale fibrous electrolytes for high-performance energy storage devices. He completed his engineering doctoral degree in less than three years, which is still an unbroken record in the Republic of Korea. After the completion of his PhD, Dr. Prasanth joined Nanyang Technological University (NTU), Singapore, as a research scientist in collaboration with the Energy Research Institute at NTU (ERI@N) and TUM CREATE, a joint electromobility research project between Germany's Technische Universität München (TUM) and NTU, where he worked with Prof. Rachid Yazami, who successfully introduced the graphitic carbon as anode for commercial lithium ion batteries. After four years in Singapore, Dr. Prasanth moved to Rice University,

Houston, Texas, USA, as a research scientist, where he worked with Prof. Pulickal M. Ajayan, the co-inventor of carbon nanotubes. Dr. Prasanth has been selected for prestigious fellowships including the Brain Korea Fellowship (2007); the SAGE Research Foundation Fellowship, Brazil (2009); the Estonian Science Foundation Fellowship; the European Science Foundation Fellowship (2010); and Faculty Recharge, UGC (2015). He has received many international awards, including the Young Scientist award, Korean Electrochemical Society (2009); and the Bharat Vikas Yuva Ratna Award (2016). He has developed several products such as a high-performance breaking parachute, flex wheels for space shuttles, and high-performance lithium ion batteries for leading portable electronic device and automobile industries. He has a general research interest in polymer synthesis and processing, nanomaterials, green/nanocomposites, and electrospinning. His current research focuses on nanoscale materials and polymer composites for printed and lightweight charge storage solutions, including high-temperature supercapacitors and batteries. He has published many high-quality research papers and book/book chapters in prestigious journals and has more than 5000 citations and an h-index of 30 plus. As well as being a scientist and technology professional, Dr. Prasanth is also a poet, an activist, and a columnist in online portals and printed media.

Stelbin Peter Figerez is a graduate student at the Department of Polymer Science and Rubber Technology, Cochin University of Science and Technology (CUSAT). He works in the area of high-performance lithium ion batteries and supercapacitors under the direction of Prof. Prasanth Raghavan. He received his B.Tech degree in Polymer Science and Engineering from CUSAT. As a project fellow, he had the opportunity to work with Prof. Jou-Hyeon Ahn at Gyeongsang National University, Republic of Korea; and with Dr. Tharangathu

Narayanan Narayanan at Tata Institute of Fundamental Research, Hyderabad. His thesis work consisted of the development of nanocomposite polymer electrolyte-based carbon nanomaterials for high-performance lithium ion batteries operating at high temperatures. In his research, he followed the novel perspective of addressing the key problems of polymer electrolytes by using polymer electrolytes based on nanofibrous membranes prepared by electrospinning. He has published full-length research articles and book chapters and has delivered invited talks at international conferences on polymer electrolytes and the nanoengineering of electrodes for the development of next-generation lithium ion batteries. Mr. Figerez has won many first prizes in poster and oral presentations at international conferences in recognition of his innovative research in the field of lithium ion batteries. His current research focuses on advanced nanomaterials and the sustainability aspects of high-temperature lithium ion batteries, including life-cycle assessment considerations, carried out in collaboration with Gyeongsang National University, Republic of Korea, and Tata Institute of Fundamental Research, Hyderabad. His current area of interest is flexible, printable, and rollable energy storage solutions and the development of novel nanomaterials for sustainable energy applications like supercapacitors, fuel cells, and lithium ion batteries.

List of Symbols and Formulas

Ag	Silver
AgCl	Silver chloride
Al	Aluminum
Al$_2$O$_3$	Alumina
B	Boron
BaTiO$_3$	Barium titanate
Bi	Bismuth
BN	Boron nitride
C	Carbon
CCl$_4$	Carbon tetrachloride
CF$_3$SO$_3$Li	Lithium trifluoromethanesulfonate
Co	Cobalt
Co$_2$SnO$_4$	Cobalt stannate
CoO	Cobalt oxide
COOH	Carboxylic acid
Cu	Copper
Fe	Iron
Fe(NO$_3$)$_3$	Ferric nitrate
Fe$_2$O$_3$	Ferric oxide (haematite)
Fe$_3$O$_4$	Ferroso ferric oxide (magnetite)
FeS	Iron(II) sulfide or ferrous sulfide
Ga	Gallium

Ge	Germanium
KMnO$_4$	Potassium permanganate
Li	Lithium
Li$_2$MnO$_3$	Dilithium,dioxido(oxo)manganese
Li$_2$O	Lithium oxide
Li$_2$O$_2$	Lithium peroxide
Li$_2$S	Lithium sulfide
Li$_3$PO$_4$	Lithium phosphate
LiClO$_4$	Lithium perchloride
LiCoO$_2$	Lithium cobalt oxide
LiFe$_5$O$_8$	Lithium ferrite
LiFeO$_2$	Lithium ferrite
LiFePO$_4$	Lithium ion phosphate
LiMn$_2$O$_4$	Lithium manganese oxide
LiMnO$_2$	Lithium manganese oxide
LiNiO$_2$	Lithium nickel dioxide
LiNO$_3$	Lithium nitrite
LiPF$_6$	Lithium hexafluorophosphate
LiTFSi	Lithium bis(trifluoromethanesulfonyl)imide
Mg	Magnesium
Mn	Manganese
MnO$_2$	Manganese dioxide
Mo	Molybdenum
MoS$_2$	Molybdenum disulphate
N	Nitrogen
Nb$_2$O$_5$	Niobium pentoxide
Ni	Nickel
NiO	Nickel monoxide
OH	Hydroxide
P	Phosphorous
Sb	Antimony
Si	Silicon
SiO$_2$	Silicon dioxide
Sn	Tin
SnCl$_2$.2H$_2$O	Tin (II) chloride dihydrate

SnO_2	Tin dioxide
SnSb	Tin antimony alloy
Ti	Titanium
TiO_2	Titanium dioxide
TiS_2	Titanium disulphide
V_2O_5	Vanadium pentoxide
VO_4	Vanadate ion
Zn_2SnO_4	Zinc stannate
ZrO_2	Zirconium dioxide
$\gamma\text{-LiAlO}_2$	Gamma phase lithium aluminate

List of Abbreviations

0D	Zero-dimension
1D	One-dimension
2D	Two-dimension
3D	Three-dimension
A-CNT	Aligned carbon nanotube
BCP	Branched-graft copolymer
C_h	Chiral vector
CNF	Carbon nanofiber
CNT	Carbon nanotube
CVD	Chemical vapor deposition
DEC	Diethyl carbonate
DHPG	Doped hierarchically porous graphene
DLG	Double later graphene
DMC	Dimethyl carbonate
DWNT	Double-walled carbon nanotube
E_a	Activation energy
EC	Ethylene carbonate
Eg	Band gap
EMC	Ethyl methyl carbonate
EO	Ethylene oxide
FET	Field-effect transistors
FLG	Few layer graphene
GNR	Graphene nanoribbon
GNS	Graphene nanosheets
GO	Graphene oxide

GPE	Gel polymer electrolyte
h-BN	Hexagonal boron nitride
h-MWNT	Hollow-multi-walled carbon nanotube
Li$^+$-ion	Lithium ion
LIB	Lithium ion battery
Li-S	Lithium-sulfur
M-PO$_4$	Metal phosphate
MWNT	Multi-walled carbon nanotube
Na-S	Sodium-sulfur
Ni-Cd	Nickel-cadmium
Ni-MH	Nickel-metal hydride
NMP	N-Methyl-2-pyrrolidone
NTCDA	1,4,5,8-naphthalenetetracarboxylic dianhydride
OLO	Over-lithiated layered oxide
ox-GNR	Oxygenated carbon nanoribbon
P(VdF-co-HFP)	Polyvinylidene difluoride co-hexafluoropropylene
PAN	Poly(acrylonitrile)
PANI	Polyaniline
PC	Propylene carbonate
PE	Polymer electrolyte
PEDOT	Poly(3,4-ethylenedioxythiophene)
PEG	Polyethylene glycol
PEGMA	Poly(ethylene glycol) methyl ether methacrylate
PEO	Poly(ethylene oxide)
PET	Polyethylene terephthalate
PGO	Polyethylene glycol grafted graphene
PMDA	Pyromellitic dianhydride
PMMA	Poly(methyl methacrylate)
PNC	Polymer nanocomposite
PPy	Polypyrrole
PVdF	Poly(vinylidene fluoride)
PVP	poly(vinyl pyrrolidone)

R_b	Bulk resistance
rGO	Reduced graphene oxide
s	Second
SANT	Super-aligned carbon nanotube
SEI	Solid electrolyte interface
SEM	Scanning electron microscope
SLG	Single layer graphene
S-PAN	Sulfur bonded polyacrylo nitrile
SPE	Solid polymer electrolyte
S-PS	Sulfonated polystyrene
Super P	Conducting carbon
SWNT	Single-walled carbon nanotube
TEM	Transmission electron microscope
T_g	Glass transition temperature
Tm	Melting temperature
t-MWNT	Thin-multi-walled carbon nanotube
VTF	Vogel–Tamman–Fulcher
α-CNT	Amorphous carbon nanotube
V_F	Fermi velocity

List of Units

°C	Degree Celsius
μm	Micrometer
A g^{-1}	Ampere per gram
Å	Armstrong
Ah kg^{-1}	Ampere hour per kilogram
cm^2 s^{-1}	Square centimeter per second
cm^2 V^{-1} s^{-1}	Square centimeter per volt per seconds
eV	Electron volt
GPa	Giga pascal
h	Hour
kcal mol^{-1}	Kilo calorie per mole
kW kg^{-1}	Kilowatt per kilogram
m s^{-1}	Meter per second
M	Molar
m^2 g^{-1}	Square meter per gram
mA g^{-1}	Milliampere per gram
mAh cm^{-2}	Milliampere hour per square centimeter
mAh cm^{-3}	Milliampere hour per cubic centimeter
mAh g^{-1}	Milliampere hour per gram
meV	Millielectron volt
mM	Millimolar
MPa	Mega pascal
MPa$\sqrt{m}$	Megapascal sqrt (meter)
mS cm^{-1}	Milli siemens per centimeter
mV	Millivolt

mWh cm^{-3}	Milliwatt hour per cubic centimeter
nm	Nanometer
Nm^{-1}	Newton per meter
nN	nanonewton
S cm^{-1}	Siemens per centimeter
TPa	Tera pascal
V	Volt
W kg^{-1}	Watts per kilogram
Wh kg^{-1}	Watt hour per kilogram
Wh L^{-1}	Watt hour per liter
Wm^{-1} K^{-1}	Watt per meter per kelvin
Ω	Ohm

CHAPTER 1

Lithium Ion Batteries

History, Basics and Challenges

1.1 INTRODUCTION

Energy plays an exceptional role in human life, and its demand and supply have always been among the crucial factors for the evolution of civilization. The demand for energy drastically increases with time or population. Energy has always been among the most essential resources for supporting the process of evolution and the prosperity of human life, for example, for heating and cooling, operating electronic and electrical appliances, transportation, communication, recreation, and so on. In this modern era, one cannot even imagine a day without using energy [1]. Global energy consumption is expected to increase by 28% by 2040 [2]. It is widely reported that future energy demand cannot be satisfied by current technologies. There are also predictions that the next world war will be over energy [3]. As we begin a revolution of automation and electric/hybrid zero-emission vehicles, we need more advanced and environmentally friendly technologies. The current reserve of fossil-fuel energy sources will be depleted in a few decades, basically due to high demand and, in some cases, overconsumption. Petroleum, natural gas, and coal are generally referred to as *fossil fuels* [1, 4].

Recently, the inevitable depletion of non-renewable fossil fuels and the move toward zero-emission vehicles and the dream of a clean environment have forced mankind to move away from using fossil fuels as the main global energy source. Green energy sources, such as solar, hydroelectric, thermal, wind, and tidal energy, will eventually replace traditional energy sources; however, most of these renewable energy sources are typically periodic or intermittent. The production of electricity using the aforementioned renewable sources needs electrochemical energy storage devices such as batteries, supercapacitors, and fuel cells, which play an important role in the efficient use of renewable energy during times of depletion. Figure 1.1 shows the Ragone plot of the specific power against specific energy for various electrochemical energy storage systems [5]. Among the various electrochemical energy storage devices available, batteries are crucial, as they can efficiently

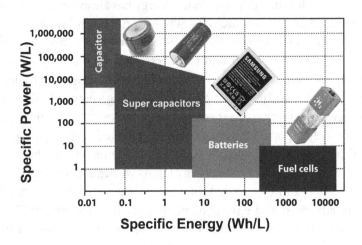

FIGURE 1.1   Ragone plot showing specific power against specific energy for various electrochemical energy storage systems.

store electricity in chemical form and release it according to demand.

A battery is a collective arrangement of electrochemical cells in which chemical energy is converted into electricity; it can be used as a source of power when a chemical reaction releases the stored the energy as electrons and ions [6]. The first battery, a Voltaic pile consisting of a series of copper and zinc discs separated by cardboard moistened with a salt solution, was developed by Alessandro Volta in the year 1800. After more than 200 years of development, battery technology has reached a point where batteries can be made in any sizes ranging from macro to nano, in shapes ranging from cylindrical to prismatic or even paper batteries, fabrication techniques from roll to roll printing to paintable batteries [7, 8] and is useful for various applications.

1.2 CLASSIFICATION OF BATTERIES

The two mainstream classifications of batteries are disposable/non-rechargeable (primary) and rechargeable (secondary) batteries. A primary battery is designed to be used once and then discarded rather than recharged with electricity. In general, primary batteries are assembled in a charged condition, and the electrochemical reaction occurring in the cell is mostly irreversible, rendering the cell non-rechargeable. Some examples of primary batteries are Lechlanche, alkaline manganese dioxide, silver oxide, and zinc/air batteries [5].

A secondary battery is a type of electrochemical device in which the chemical reactions can be reversed by an external electrical energy source. Therefore, these cells can be recharged numerous times by passing an electric current through them after they have reached their fully discharged state and used it for long periods. Generally, secondary batteries have a lower capacity and initial voltage, a flat discharge curve, higher self-discharge rates, and varying recharge life ratings. Secondary batteries usually have more active (less stable) chemistries, which need special handling, containment, and disposal. Some examples of secondary batteries are aluminum ion, lead acid, lithium ion, nickel cadmium, and sodium ion batteries. Depending on the chemical reaction involved, rechargeable batteries can further be classified as lead-acid, nickel-metal hydride, zinc air, sodium-sulfur, nickelcadmium, lithium ion, lithium air, and so on. Batteries may also be classified by the type of electrolyte employed, either an aqueous or a non-aqueous system. Some common battery types are shown in Table 1.1, and the characteristics and performance of commonly used rechargeable batteries are tabulated in Table 1.2 in

TABLE 1.1   Common Battery Types Based on the Type of Materials and Their Rechargeability

| Type of Battery | Type of Electrolyte in the Battery | |
	Aqueous Electrolyte (Low Voltage Capacity)	Non-aqueous Electrolyte (High Voltage Capacity)
Primary battery (disposable)	Manganese dry cell Alkaline dry cell Li-air battery	Li-metal battery
Secondary battery (rechargeable)	Lead-acid battery Ni-Cd battery Ni-MH battery Sodium ion battery	Al-ion battery LIB Li-air battery

TABLE 1.2   The Characteristics and Performance of Commonly Used Rechargeable Batteries

Battery Type	Lead-Acid	Ni-Cd	Ni-MH	Li^+-Ion
Commercialization (year)	1970	1956	1990	1992
Nominal cell voltage (V)	2.1	1.2	1.2	3.6
Volumetric energy density (Wh L^{-1})	60–75	50–150	140–300	250–620
Gravimetric energy density (Wh kg^{-1})	30–50	40–60	60–120	100–250
Power density (W kg^{-1})	180	150	250–1000	250–340
Cycling stability	500–800	2000	500–1000	400–1200
Monthly self-discharge rate (%) at room temperature	3–20	10	30	5
Memory effect	No	Yes	No	No

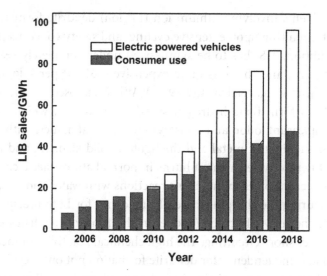

FIGURE 1.2 Forecasted expansion in demand for lithium ion batteries from 2005 to 2018 in electric vehicles and consumer applications.

accordance with these classifications. Among the aforementioned rechargeable batteries, lithium ion batteries (LIBs) have gained considerable interest in recent years in terms of the highest specific energy they can store, their cell voltage, their good capacity retention, and their negligibly small self-discharge [6]. Figure 1.2 shows the forecasted expansion in demand for LIBs in the current decade, and it is evident that the importance of LIBs in day-to-day life is greatly increasing over time.

1.3 HISTORY OF LITHIUM ION BATTERIES

The brief history of the development of LIBs is quite interesting. The first rechargeable LIBs were reported by the British chemist M. Stanley Whittingham, a key figure in the history of the development of LIBs, while he was working at Exxon. These batteries were fabricated in 1976 with layered titanium disulfide (TiS_2) as the cathode, and metallic lithium as an anode [9]. Exxon subsequently tried to commercialize the LIBs; however, this failed due to problems involving lithium ion (Li+-ion) dendrite formation, short-circuiting upon extensive cycling, and safety concerns [10]. In addition, TiS_2 has to be synthesized under completely sealed conditions, and this was quite expensive (~$1000 per kilogram for TiS_2 raw material in the 1970s). When exposed to air, TiS_2 reacts to form toxic hydrogen sulfide compounds, which have an unpleasant odor and cause environmental issues. Lithium (Li) is the lightest metal and the lightest solid element, and it is a highly reactive element; it burns in normal atmospheric conditions because of its spontaneous reactions with water and oxygen [11]. During the charging cycle, the tendency for Li to precipitate on the negative electrode (anode) in the form of dendrites easily causes short-circuiting. The high chemical reactivity of metallic Li and the tendency for dendrite formation not only results in poor battery characteristics, including inadequate cycling stability because of side reactions with the electrolyte, but also poses an inherent risk of a thermal runaway reaction, which was an insurmountable issue in terms of safety. As a result, researchers focused on developing LIBs that employed only Li compounds, capable of accepting and releasing Li+-ions, instead of metallic Li electrodes. As a result, reversibly intercalating Li+-ions into graphite [12–14] and cathodic oxides [15–17] was reported by J. O. Besenhard in 1976, suggesting its application as anode and cathode in Li+-ion cells [14, 17].

In 1977, Samar Basu demonstrated electrochemical intercalation of Li+-ion in graphite, which led to the development at Bell Labs of a workable Li+-ion intercalated graphite electrode (LiC_6) as an alternative to the Li metal battery [18, 19]. In 1979 and 1980, respectively, Ned A. Godshall et al. [20–22] and John Goodenough et al. [23–25] demonstrated a rechargeable Li+-ion cell with nominal voltage of 4 V, using layered $LiCoO_2$ as the high-energy and high-voltage material for the positive electrode and Li metal as the negative electrode; however, layered $LiCoO_2$ did not attract much attention initially. In 1979, Godshall et al. demonstrated that other ternary compound Li-transition metal oxides, such as the spinel

$LiMn_2O_4$, Li_2MnO_3, $LiMnO_2$, $LiFeO_2$, $LiFe_5O_8$, and $LiFe_5O_4$, could be used as alternatives to $LiCoO_2$ for electrode materials for LIBs [20–22, 26], and they were awarded a US patent in 1982 on the use of $LiCoO_2$ as cathodes in LIBs [27]. In 1983, Goodenough also identified manganese spinel as a low-cost cathode material [28], and in 1985, Godshall et al. identified Li-copper-oxide and Li-nickel-oxide cathode materials for LIBs [26]. The lack of safe anode materials, however, limited the application of the layered oxide cathode $LiMO_2$ (M = Mn, Ni, Co) in LIBs.

In 1978 and 1980, respectively, Besenhard [14] and Basu [19] and Rachid Yazami [29, 30] demonstrated that graphite, also with a layered structure, could be a good candidate for reversibly storing Li^+-ion by electrochemical intercalation/deintercalation of Li^+-ion in graphite. The publications of Yazami and Touzain are accepted as the world's first successful experiments demonstrating the electrochemical intercalation and release of Li^+-ion in graphite. The organic electrolytes available for LIBs at the time would decompose when the cell was charged using graphite as an anode, slowing the commercialization of a practical rechargeable Li/graphite battery. In Yazami's studies, a solid electrolyte is used to demonstrate that Li could be reversibly intercalated in graphite through an electrochemical mechanism, and this experiment provided the scientific basis for the use of graphite as negative-electrode material, as is standard in LIBs today. (As of 2011, the graphite electrode discovered by Yazami is the most commonly used electrode in commercial LIBs.)

In 1985, Akira Yoshino assembled a prototype Li^+-ion cell using a carbonaceous anode (polyacetylene, which is an electrically conductive polymer discovered by Professor Shirakawa, who received the Nobel Prize for chemistry in 2000), into which Li^+-ions could be inserted and discharged; $LiCoO_2$, which is stable in air, was used as the cathode [31, 32]. Both the carbon anode and $LiCoO_2$ cathode are stable in air, which is highly beneficial from engineering and manufacturing perspectives. Also, Yoshino [33] found that carbonaceous material with a certain crystalline structure provided greater capacity without causing decomposition of the propylene carbonate electrolyte solvent, as graphite electrode did. This battery design, using materials without metallic Li, enabled industrial-scale manufacturing and heralded the birth of the current LIB. After the successful demonstration of this prototype design for the LIB, in 1986, Yoshino carried out the world's first safety tests on LIBs and proved that this LIB overcame the safety issues that had previously prevented the commercialization of non-aqueous secondary batteries. Because of the risk of ignition or even explosion during the safety test, Yoshino had to borrow a facility designed for testing explosives. In these tests, an iron lump was dropped onto the batteries and compared with a set of cells assembled using a Li metal electrode; the test result showed that violent ignition occurred with a metallic Li battery, while no ignition occurred with the LIB. According to Yoshino, this was a great relief, because if ignition had occurred during this test, the LIB would not have been commercialized. This was the crucial turning point for the commercialization of the LIB. I consider the success of these tests to be "the moment when the lithium ion battery was born" [33].

Eventually, Sony, the dominant maker of personal electronic devices such as the Walkman and photographic cameras at that time, commercialized LIBs in 1991 and by a joint venture of Asahi Kasei and Toshiba in 1992. Table 1.3 shows some milestones in commercialization of LIBs. It was a tremendous success and facilitated a major reduction in the size and weight of power supplies for portable devices, thereby leading to the revolution of the portable electronics industry. Commercialization of the LIB made available an energy density that, in terms of both weight and volume, was around twice that is possible with nickel cadmium or nickel-metal hydride batteries; moreover, by providing an electromotive force of 4 V or more, the LIB made it possible to power a cell phone with a single cell. To acknowledge their pioneering contribution to the development of LIB, Goodenough, Yazami, and Yoshino were awarded the 2012 IEEE Medal for Environmental and Safety Technologies and the Draper Prize in 2014.

TABLE 1.3   Historical Milestones of the Commercial Lithium Ion Batteries

Year	Industry	Historical Milestone
1991	Sony (Japan)	Commercialization of LIB
1994	Bellcore (USA)	Commercialization of Li polymer
1995	Group effort	Introduction of pouch cell using Li polymer
1995	Duracell and Intel	Proposal of industry standard for SMBus[a]
1996	Moli Energy (Canada)	Introduction of Li+-ion with manganese cathode
1996	University of Texas (USA)	Identification of Li phosphate
2002	Group effort	Various patents filed on nanomaterials for batteries

[a] System management bus.

1.4 PRINCIPLE OF LITHIUM ION BATTERIES

A primary LIB is a one-direction device that only has a discharging process. During discharging, there is a reduction in the cathode gaining electrons and oxidation reacts on the anode losing electrons, displayed in following reaction [6].

$$\text{Cathode:}\quad MS_2 + Li^+ + e^- \xrightarrow{\text{discharge}} LiMS_2$$

$$\text{Anode:}\quad Li \xrightarrow{\text{discharge}} Li^+ + e^-$$

$$\text{Full cell:}\quad Li + MS_2 \xrightarrow{\text{discharge}} LiMS_2 \quad \left(M = Ti \text{ or } Mo\right)$$

In contrast to primary cell batteries, secondary cell LIBs are rechargeable. Figure 1.3 shows the schematic representation of the working principle of LIBs. During the charge/discharge process, the oxidation and reduction process occurs at two electrodes, as shown here.

$$\text{Cathode:}\quad LiMn_2O_4 \rightleftarrows Li_{1-x} + xMn_2O_4 + xLi^+ + xe^-$$

$$\text{Anode:}\quad xLi^+ + xe^- + C_6 \rightleftarrows Li_xC_6$$

$$\text{Full Cell:}\quad LiMn_2O_4 + C_6 \rightleftarrows LiC_6 + LiMn_2O_4$$

The secondary LIBs, in general, operate at 3.7 V and demonstrate a capacity of 150 mAh g^{-1}.

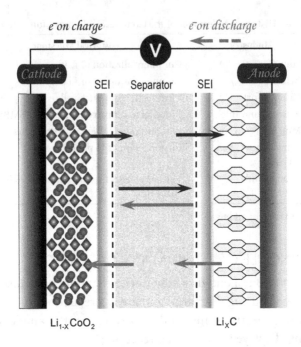

FIGURE 1.3 Schematic illustration of operating principles of LIBs including the movement of ions between electrodes during charge (dark arrow) and discharge (light arrow) states.

1.5 CHALLENGES FOR NEXT-GENERATION LITHIUM ION BATTERIES

As discussed at the beginning of this chapter, a move toward the electrification of road transportation is becoming a societal goal of vital importance. Among other factors such as finite fossil-fuel supplies and a need to curb global warming, issues associated with air quality have become pressing, making a major push toward the adoption of electric vehicles in densely populated cities in order to improve the quality of breathable air a worldwide imperative. As a mature technology, LIBs have aided the revolution in microelectronics and have become the power source of choice for portable electronic devices. Their triumph in the portable electronics market is due to the higher gravimetric and volumetric energy densities offered by LIBs compared to other rechargeable systems; however, for powering electric vehicles, LIBs are still unfeasible. A quote featured in 1915 by the *Washington Post* summarized well the current situation of electric vehicles: "Prices on electric cars will continue to drop until they're within the reach of the average family" [34]. Even a century later, the automobile industry is still facing the same cost issue. For portable electronics, energy density is the most important factor, and current LIBs are sufficient to power portable electronic devices; however, for electric vehicles, power density, cost, cycle life, and safety also become critical performance parameters along with energy density (driving distance between charges). Adoption of electric vehicles is limited primarily due to the cost, safety issues, and inadequate storage capacity of today's LIBs.

The performance parameters of LIBs are largely determined by the electrochemical properties and characteristics of the component materials used in fabricating the batteries, as well as the cell engineering and system integration involved. The characteristics of the materials employed rely on the underlying chemistry associated with the materials. Among current battery technologies, Li^+-ion technology performs best owing to its gravimetric energy densities of <250 Wh kg^{-1} and volumetric energy densities of <650 Wh L^{-1}, which exceed any competing technologies by at least a factor of 2.5 [35]. However, the adoption of LIBs for powering electric vehicles represents a colossal task. If we really want to compete with gasoline, an increase by a factor of 12 is needed if the energy delivered by a battery is to match that of a liter of gasoline (3000 Wh L^{-1}; taking into account corrections from Carnot's principle). Knowing that the energy density of batteries has only increased by a factor of 5 over the last two centuries, our chances of achieving a 10-fold increase over the next few years are very slim, in the absence of unexpected research breakthroughs. Fortunately, the automotive industry has set a more realistic target: doubling the present Li^+-ion energy density (as high as ~500 Wh kg^{-1} and >1,000 Wh L^{-1}) over the next 10 years so that the autonomy of electric vehicles approaches 500 km. Accomplishing this goal is challenging; it will require innovations both in the component materials used in the cell and in the engineering involved in fabricating the cells. It should be recognized that the incremental improvements made in energy density since the first announcement of commercial LIBs in 1991 by the Sony Corporation have largely been due to progress in terms of engineering, as the component electrode materials remain the same, with minor modifications.

The energy density of a battery is the product of its capacity and its potential, and it is mainly governed by the capacity of the positive electrode. Simple calculations show that an increase in cell energy density of 57% can be achieved by doubling the capacity of the positive electrode, while the capacity of the negative electrode needs to be increased by a factor of 10 to get an overall cell energy density increase of 47% [36]. So the chances of drastically improving the energy density of today's Li^+-ion cells are mainly rooted in spotting better positive electrode materials, that is, materials that could either display greater redox potential (e.g. are highly oxidizing) or larger capacity (materials capable of reversibly inserting more than one electron per 3d metal.)

Many next-generation conversion reaction cathodes such as sulfur (or Li_2S) and oxygen (or Li_2O_2 or Li_2O) and anodes such as Si, Sn, Sb, Ge, and P are being actively pursued. The conversion reaction anodes offer much higher capacities than graphite, but they have higher operating voltages than graphite (which would lower the cell voltage) and suffer from large volume expansion and contraction (up to ~400% depending on the anode and the Li content, compared to <10% for graphite) during charging and discharging, respectively [37]. The large volume changes during the charge/discharge cycling, pulverization of the particles, continuous formation of solid electrolyte interface (SEI), and consequent trapping of active Li from the cathode in the anode SEI [38] are the major challenges posed by conversion reaction anodes. Many approaches have been pursued, such as reducing the particle size from micro to nano size, making composite electrodes with carbon nanotube (CNT) or graphene, and deliberately leaving space within the active material architecture, but none of them are as yet successful enough to be practically viable [39, 40]. The above approaches drastically increase SEI formation and decrease the volumetric energy density. The particle milling caused by volume changes results in the continuous formation of new surfaces during the charge/discharge process, which further aggravates the formation of SEI, posing daunting challenges. It is well known that the stable and reliable functioning of LIBs hinges on the formation of a stable electrode–electrolyte interface layer, which is conductive to Li^+-ion while electronically insulating. The importance of the SEI layers for battery performance is widely acknowledged; however, despite over four decades of research, the exact mechanisms of the formation of SEI layers still remain poorly understood, and understanding of the composition and properties of the SEI layer is limited. For the development of next-generation LIBs, in-depth understanding of the reactions with the electrolyte in light of recent developments in high-capacity positive electrodes is critically needed [41]. The concept of developing a systematic framework that links electrochemistry and solid mechanics for LIBs remains in its infancy. In this scenario, with the aim of resolving the related issues, a recent report discussed in detail the electrochemical–mechanical coupling and its properties in the context of ion conducting soft materials. The report also provides a comprehensive discussion of the various types of defects that can form in solid electrolyte membranes or at their interfaces [42]. Even though SPEs are a well-studied electrolyte system for the development of safer batteries, the development of a practical SPE that has good electrochemical performance and thermomechanical and dimensional stability without compromising its Li^+-ion transport resistance and transport number still remains a big challenge.

Sulfur and oxygen (the so-called "Beyond Li^+-ion" chemistries) offer much higher capacities than layered, spinel, and olivine-type cathodes. Oxygen-based cathodes suffer from clogging by insoluble products, catalytic decomposition of electrolytes, moisture from the air, and poor cycle life, making their practical viability extremely difficult, if not impossible. The critical challenges of sulfur-based cathodes are much less than those of oxygen-based cathodes, and in recent years remarkable progress has been made in increasing the active material content and loading, suppressing dissolved polysulfide migration between the electrodes, and reducing the amount of electrolyte [43, 44]. However, the necessity of pairing a Li metal anode with a sulfur or oxygen cathode poses formidable challenges, unless Li_2S and Li_2O_2 cathodes that can be successfully paired with an anode like graphite or Si or practical lithium-containing anodes that could be paired with sulfur or oxygen can be developed. On the basis of simple Faradaic calculations, it is reported that an electrolyte-to-sulfur ratio of 11.1 and 4.0 represents the point where an "idealized" lithium sulfur (Li-S) battery will surpass a traditional LIB in terms of a specific energy (Wh kg^{-1}) and energy density (Wh L^{-1}). Based on close data analysis, McCloskey makes a strong recommendation that all future Li-S battery studies focus on an electrolyte-to-sulfur ratio of less than 11 and more realistically of about 5. The literature shows that a practical Li-S battery can compete favorably on volumetric energy density with an LIB comprising a Li metal anode coupled with an advanced intercalation cathode material. To suppress dendrite formation and reduce parasitic reactions between the Li and electrolyte constituents, the practical Li-S battery requires a solid-state electrolyte or separator. To make the Li-S cell a cost-competitive alternative to LIBs, a cost target of ~$10 m^{-2} is identified for the solid-state separator [45]. Practically, these targets are daunting challenges for battery technologists and require innovative solutions. In a nutshell, the challenges for the development of next-generation LIBs primarily focus on the following directions: increasing energy density and safety, reducing cost, achieving sustainable and greener LIBs, and increasing capacity while ensuring sustainability and green storage.

1.6 SUMMARY

LIBs have clear fundamental advantages, and decades of research history have gone into developing today's LIBs, which have a high energy density, power density, cycling stability, rate capability, and columbic efficiency. As energy demand is rocketing day by day, LIB research is continuing on high-performance batteries that can be used as efficiently even in abused conditions. Because of the global effort to establish zero-emission automobiles, battery technologists are dedicated to finding new electrode materials to push the boundaries of cost, energy density, power density, cycle life, rate capability, and safety of LIBs for heavy-duty applications such as electric vehicles. This chapter narrates the history of the birth of LIBs and the important milestones in their development up to the present, as well as paying tribute to the eminent personalities responsible for the development and commercialization of practical LIBs. The chapter also offers an insight into the challenges and opportunities of next-generation LIBs.

REFERENCES

1. https://www.conserve-energy-future.com/energyconsumption.php
2. US Energy Information Administration—International Energy Outlook 2017.
3. https://oilprice.com/Energy/Energy-General/World-War-III-Its-Here-And-Energy-Is-Largely-Behind-It.html
4. Kumar et al. *Renew. Sustain. Energy Rev.* 14 (2010) 2434–2442.
5. Winter et al. *Chem. Rev.* 104 (2004) 4245.
6. Lim et al. *J. Nanosci. Nanotechnol.* 18 (2018) 6499.
7. Singh et al. *Sci. Rep.* 2 (2012) 481.
8. Hu et al. *ACS Nano* 4 (2010) 5843.
9. Whittingham. *Science* 192 (1976) 1126.
10. Levine. *Foreign Policy* 182 (2010) 88.
11. Kirchhoff et al. *Q. J. Chem. Soc.* 13 (1861) 270.
12. Besenhard et al. *J. Electroanal. Chem.* 53 (1974) 329.
13. Besenhard. *Carbon* 14 (1976) 111.
14. Besenhard et al. *J. Electroanal. Chem. Interfacial Electrochem.* 68 (1976) 1.

15. Schöllhorn et al. *Mater. Res. Bull.* 11(1976) 83.
16. Besenhard et al. *J. Power Sources* 1 (1976) 267.
17. Eichinger et al. *J. Electroanal. Chem. Interfacial Electrochem.* 72 (1976) 1.
18. Zanini et al. *Carbon* 16 (1978) 211.
19. Basu et al. *Mater. Sci. Eng.* 38 (1979) 275.
20. Godshall et al. *Mater. Res. Bull.* 15 (1980) 561.
21. Godshall N.A. (17 October 1979) Electrochemical and thermodynamic investigation of ternary lithium-transition metal-oxide cathode materials for lithium batteries: Li_2MnO_4 spinel, $LiCoO_2$, and $LiFeO_2$. Presentation at 156th Meeting of the Electrochemical Society, Los Angeles, CA.
22. Godshall N.A. (18 May 1980) Electrochemical and thermodynamic investigation of ternary lithium-transition metal-oxygen cathode materials for lithium batteries. Ph.D. Dissertation, Stanford University.
23. Mizushima et al. *Mater. Res. Bull.* 15 (1980) 783.
24. Mizushima et al. *Solid State Ionics* 3–4 (1981) 171–174.
25. Goodenough et al. EP17400B1, 1979.
26. Godshall et al. *Solid State Ionics.* 18–19 (1986) 788.
27. Godshall et al. "Ternary Compound Electrode for Lithium Cells"; issued 20 July 1982, filed by Stanford University on 30 July 1980. US Patent 4,340,652.
28. Thackeray et al. *Mater. Res. Bull.* 18 (1983) 461.
29. International Meeting on Lithium Batteries, Rome, 27–29 April 1982, C.L.U.P. Ed. Milan, Abstract #23.
30. Yazami et al. *J. Power Sources* 9 (1983) 365.
31. Yoshino et al. Japanese Patent Application 1985-127669, 1985.
32. Yoshino et al. USP4,668,595, 1987.
33. Yoshino. The birth of the lithium-ion battery. *Angew. Chem. Int.* 51 (2012) 5798–5800.
34. http://www.nationalreview.com < A Million Electric Vehicles (June 7, 2011)
35. Tarascon et al. *Nature* 414 (2001) 359.
36. Tarascon et al. *Act. Chim.* 251 (2002) 130.
37. Li et al. *Nat. Commun.* 5 (2014) 4105.
38. Pan et al. *Chem. Commun.* 50 (2014) 5878.
39. Chan et al. *Nat. Nanotechnol.* 3 (2008) 31.
40. Yoon et al. *Chem. Mater.* 21 (2009) 3898.
41. Gauthier et al. *Phys. Chem. Lett.* 6 (2015) 4653.

42. Kusoglu et al. *Phys. Chem. Lett.* 6 (2015) 4547.
43. Manthiram et al. *Chem. Rev.* 114 (2014) 11751.
44. Chung et al. *ACS Nano* 10 (2016) 10462.
45. McCloskey. *Phys. Chem. Lett.* 6 (2015) 4581.

CHAPTER 2

Carbon Nanomaterials Are Resolving the Challenges and Issues of Future Lithium Ion Batteries

2.1 INTRODUCTION

The right selection of materials plays a key role in the development of advanced LIBs. To date a huge number of materials ranging from graphitic carbon to metal oxides are tested for the development of high-performance lithium ion batteries (LIBs). The first commercial rechargeable LIBs demonstrated by Whittingham et al. [1] utilized layered TiS_2 as the cathode and metallic Li as an anode, where TiS_2 serves as a host for incorporating reversible intercalated and de-intercalated Li into its structure [1]. The single-phase behavior during cycling enables it to fully remove and insert Li^+-ions reversibly; however, the formation and subsequent growth of dendrite during charge/discharge cycling can short out the cell, which will potentially lead to explosion hazards, making the commercialization of LIBs a nightmare [2]. Because of the disadvantages associated with metallic Li and TiS_2, the researchers focused on identifying economically viable and environmentally friendly alternative materials to develop practical LIBs. As a result, reversibly intercalating Li^+-ions into graphite [3–5] and cathodic oxides (Li compounds) [6–8], capable of accepting and releasing Li^+-ions, was proposed application as anode and cathode respectively in Li^+-ion cells [5, 8], instead of metallic Li electrodes. In the late 1970s, Besenhard [5], Basu [9], and Rachid Yazami [10, 11] demonstrated that graphite, which also has a layered structure, could serve as a highly reversible and low-voltage Li^+-ion intercalation/deintercalation anode, as it possesses unique properties. The publications of Yazami and Touzain are accepted as the world's first successful experiments demonstrating the electrochemical intercalation and release of Li in graphite. In graphitic carbons, Li^+-ions intercalate between graphene layers in stages that result in a maximum configuration of one Li atom to every six carbon atoms.

Combining safety features of the carbon anode and the high-voltage $LiCoO_2$ cathode, Sony commercialized the first LIBs ($C/LiCoO_2$ cell). Although graphitic anodes have the advantage of a long cycle life, low cost, environmental friendliness, and abundance, they have significant disadvantages with regards to low gravimetric and volumetric specific capacity (372 mAh g^{-1} and 833 mAh cm^{-3}). In order to improve the specific energy density (Wh kg^{-1}) in batteries, one can increase the overall battery capacity (Ah kg^{-1}) and/or V. Since the commercial LIBs employ graphite as the anode, which has a constant potential of between 0.1 and 0.2 V versus Li/Li^+, the nominal battery voltage and the Li storage capacity are highly influenced by the cathode chemistry, stoichiometry, and crystal structure. Thus, the specific energy density of the battery can be achieved by improving the cathode capacity and/or by increasing the voltage of the cathode versus Li/Li^+ [2, 12]. The other strategy that can be used to increase the energy density is to augment the anode capacity sufficiently to increase the number of active layers contained within an individual battery. Apart from the use of carbon materials as the anode, it is used to make conducting network in the cathode, promoters for Li^+-ion conduction and current collectors for flexible or paintable batteries. There are hundreds of commercially available carbon types, including natural and synthetic graphite, carbon blacks, active carbons, carbon fibers, cokes, and various other carbonaceous materials produced by the pyrolysis of organic precursors in an inert gas environment. The carbon anode materials are generally classified into three categories: (i) graphite; (ii) non-graphitized glass-like carbon (hard carbon), which cannot be graphitized even when heat treated at high-temperature; and (iii) soft carbon, easily changeable with heat treatment.

The limitations of bulk carbonaceous battery materials can be addressed by the size reduction of the particles to the nanoscale or through the introduction of other functional materials such as carbon nanotubes (CNTs) or graphene. It has been demonstrated previously that reduction of a particle size to the nanoscale (i.e. <100 nm) can change the crystal structure, potentially modifying the mechanism of volumetric expansion upon lithiation and drastically increasing the active sites for the electrochemical reaction [13–15]. Also, with the introduction of carbon nanostructures into the high-capacity metal oxide or semiconductor materials, the carbon nanostructures establish an electrically conductive percolation network that facilitates rapid electron transfer and relies upon shorter Li diffusion lengths within the battery components, resulting in higher cycling stability and rate capability [13]. An additional opportunity for the use of these carbonaceous nanostructures, in conjunction with the conventional graphitic anode or metal oxides or Li compounds, is the ability to form free-standing electrodes [14, 15]. Among different carbonaceous materials, CNT and graphene are extensively studied as electrode materials and additives for polymer electrolytes for LIBs due to their unique structure and properties.

2.2 CLASSIFICATIONS OF CARBON NANOMATERIALS

2.2.1 Allotropes of Carbon

Carbon is well known to form distinct solid-state allotropes with diverse structures and properties ranging from sp^3 hybridized diamond to sp^2 hybridized graphite. Historically, chemists have known only two allotropes, or pure forms, of carbon: graphite, a greasy, electrically conducting black substance; and diamond, crystal clear, electrically insulating material that is harder than any other solid. But they have constantly theorized about other possible carbon allotropes. There are eight main allotropes of carbon: (i) diamond, (ii) graphite, (iii) lonsdaleite, (iv) C_{60} (buckminsterfullerene or buckyball), (v) C_{540}, (vi) C_{70}, (vii) amorphous carbon, and (viii) CNT or buckytube. Until the 1960s, when "new carbon" materials were synthesized, only two allotropic forms of carbon were known, graphite and diamond, including their polymorphic modifications. Recently, "amorphous carbon" has come to be considered as the third allotrope of carbon. Graphite is the most common allotrope of carbon, the most thermodynamically stable form of carbon at room temperature. Therefore, it is used in thermochemistry as the standard state for describing the heat produced in the formation of carbon compounds. Graphite consists of a layered two-dimensional (2D) structure where each layer possesses a hexagonal honeycomb structure of sp^2 bonded carbon atoms with a C-C bond length of 1.42 Å. These single atom thick layers (i.e., graphene layers) interact via non-covalent Van der Waals forces with an interlayer spacing of 3.35 Å. The weak interlayer bonding in graphite implies that single graphene layers can be exfoliated via mechanical or chemical methods, as will be outlined in detail here. Graphene is often viewed as the two-dimensional (2D) building block of other sp^2 hybridized carbon nanomaterials in that it can be conceptually rolled or distorted to form CNTs and fullerenes. Graphite is an electrical conductor and is applicable in electronics. Graphite conducts electricity due to the delocalization of the π-bond electrons above and below the planes of the carbon atoms. These electrons are free to move and so are capable of conducting electricity. However, the electricity is only conducted along the plane of the layers. In diamond, all four outer electrons of each carbon atom are localized between the atoms in covalent bonding. The movement of electrons is constrained, and diamond does not conduct an electric current. In graphite, each carbon atom uses only three of its four outer energy level electrons in covalently bonding to three other carbon atoms in a plane. Each carbon atom contributes one electron to a delocalized system of electrons that is also a part of the chemical bonding. The delocalized electrons are free to move throughout the plane. So, graphite conducts electricity along the planes of carbon atoms. Amorphous carbon is the carbon that does not have any crystalline structure. As with all glassy materials, some short-range order can be observed, but there is no long-range pattern of atomic positions. While completely amorphous carbon can be produced, most examples contain microscopic crystals of graphite-like or even diamond-like carbon.

There are many ways to classify carbon materials based on qualities such as state, structure, and dimension. In principle, different approaches can be used to classify carbon nanostructures; however, the appropriate classification scheme depends on the field of application of the nanostructures. Carbon nanomaterials are mostly classified on the basis of their dimensionality. By this classification, the entire range of dimensionalities is represented in the nanocarbon world, beginning with zero-dimension (0D) structures (fullerenes, diamond clusters) and includes one-dimensional (1D) structures (nanotubes), 2D structures (graphene), and three-dimensional (3D) structures (nanocrystalline diamond, fullerite). A different approach is classification based on the scale of characteristic sizes of the nanomaterials, and this scheme of classification naturally allows more consideration of the complicated hierarchical structures of carbon materials, such as carbon fibers and carbon polyhedral particles. Classification based on different shapes and spatial arrangements of elemental structural units of carbon caged nanostructures also provides a very clear and useful picture of the numerous forms of carbon structures at the nanoscale [16]. Regarding the last approach, the spatial distribution of penta- and hexa-rings within structures also can provide a basis for classification [17].

In terms of a more fundamental basis for the classification of carbon nanostructures, it would be logical to develop a classification scheme based on existing carbon allotropes that is inherently connected with the nature of bonding in carbon materials. Ironically, there is no consensus at present on how many carbon allotropes/forms are defined. From time to time, publications appear proposing new crystalline forms or allotropic modifications of carbon. Whether fullerenes or carbynes are considered as new carbon allotropes depends to a large extent on the corresponding scientific advancement in the carbon materials [18, 19]. Sometimes, the "fullerene community" appears to ignore the carbines, which were discovered in the 1960s, and similarly, the "carbyne community" does not classify fullerenes as an allotrope [18]. As discussed earlier, elemental carbon exists in three bonding states corresponding to sp^3, sp^2, and sp hybridization of the atomic orbitals, and the corresponding three carbon allotropes with an integer degree of carbon-bond hybridization are diamond, graphite, and carbine [20]. All other forms of carbon are classified based on the transitional forms as mixed short-range order carbon forms and intermediate carbon forms with a non-integer degree of carbon-bond hybridization, sp^n. Mixed short-range order carbon forms, such as diamond-like carbon, vitreous carbon, soot, and carbon blacks, have more or less arranged carbon atoms of different hybridization states. Numerous hypothetical structures like graphene and "superdiamond" also come under the category of mixed short-range order carbon forms. In intermediate carbon forms with a non-integer degree of carbon-bond hybridization, sp^n have many subgroups depends on the value of integer "n" in sp^n carbon-bond hybridization. When the value of n in sp^n carbon-bond hybridization, $1 < n < 2$ includes various monocyclic carbon structures. Similarly, when $2 < n < 3$, the intermediate carbon forms comprise closed-shell carbon structures such as fullerenes (the degree of hybridization in C_{60} is ~2.28), carbon onions and nanotubes, hypothetical tori, and so on.

2.2.2 Carbon Nanotubes

Though the first report on CNTs came in 1952 by Radushkevich and Lukyanovich as hollow graphitic fibers and became in limelight in the last couple of decades due to its interesting electronic and mechanical properties [21]. The invention of CNTs, which are long, thin cylinders of carbon, was scientifically reported by Sumio Iijima in 1991 as a byproduct of fullerene synthesis and garnered him the Nobel Prize. Remarkable progress has been made in this area over two decades and the topic is still hot as far as research in nanoscience is concerned. CNTs are large macromolecules that are unique for their size, shape, and remarkable physical properties. CNTs are considered allotropes of carbon, which has a cylindrical and tubular structure; however, they are different from other allotropes of carbon like graphite, diamond, and fullerene. These are 1D structures with very high aspect ratio (length-to-diameter ratio), which opened up a new area of study in nanotechnology [22, 23].

CNTs can be thought of as sheets of graphite (a hexagonal lattice of carbon) rolled into a cylinder shape structure with a diameter in the order of a few nanometers, while their length can be of the order of several millimeters and their length-to-diameter ratio up to 132,000,000:1 [24] significantly larger than any other material. These intriguing structures have sparked much excitement in recent years, and a large amount of research has been dedicated to their understanding. These cylindrical carbon molecules have unusual properties, which are valuable for nanotechnology, electronics, optics, and other fields of materials science and technology. In particular, owing to their extraordinary thermal conductivity and mechanical and electrical properties, CNTs find applications as additives to various structural materials. For instance, nanotubes form a tiny portion of the material(s) in some (primarily carbon fiber) baseball bats, golf clubs, or car parts [25].

CNTs, tubular carbon molecules, are basically members of the fullerene structural family, but while fullerene molecules form a spherical shape, nanotubes are cylindrical structures with the ends covered by half a fullerene molecule. Their name is derived from their long, hollow structure, the walls of which are formed by one-atom-thick sheets of carbon called graphene. These sheets are rolled at specific and discrete (chiral) angles, and the combination of the rolling angle and radius decides the nanotube properties, for example, whether the individual nanotube shell is metal or semiconductor (Figure 2.1). Based on the chirality, the nanotubes are classified as metallic, non-metallic, or semiconducting

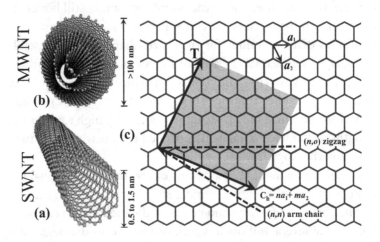

FIGURE 2.1 Schematic illustration of (a) SWNT, (b) MWNT, and (c) their unit vectors. The (n,m) nanotube naming scheme can be thought of as a vector (C_h) in an infinite graphene sheet that describes how to "roll up" the graphene sheet to make the nanotube. T denotes the tube axis, and a_1 and a_2 are the unit vectors of graphene in real space.

in nature. The way the graphene nanosheet (GNS) is wrapped can be represented by a pair of indices (n,m), where n and m denote the number of unit vectors along the two directions of a hexagonal crystal lattice of graphene. The chirality is defined by the chiral vector, $C_h = na_1 + ma_2$. CNTs have a zigzag structure for $m = 0$ and an armchair structure for $n = m$, and if both conditions are not satisfied, it is classified as chiral [26]. The chirality greatly affects the electronic properties of CNTs. For a given (n,m), if $(2n + m)$ is multiple of three, it is considered as metallic; otherwise, it is considered a semiconductor.

Several types of nanotubes exist, but they can be divided into two main categories: (i) single-walled CNTs (SWNTs)—these can be envisioned as cylinders composed of a rolled-up GNS around a central hollow core; and (ii) multi-walled CNTs (MWNTs)—these consist of two or more graphene layers held together by Van der Waals forces between adjacent layers and folded as hollow cylinders [27]. The form of the nanotube is identified by a sequence of two numbers; the first represents the number of carbon atoms around the tube, while the second identifies an offset of where the nanotube wraps around to. The chemical bonding of CNTs is sp^2 bonding similar to graphite. Based on the rolling angle of the graphite sheet against the tube axis, CNTs can have three kinds of structures: (a) armchair, (b) zigzag, and (c) chiral. Individual nanotubes naturally align themselves into "ropes" held together by Van der Waals forces, more specifically, π-stacking. Applied quantum chemistry, specifically orbital hybridization, best describes chemical bonding in nanotubes. The chemical bonds between carbon atoms inside nanotubes is always of an sp^2 type, similar to those found in graphite. These bonds, which are stronger than the sp^3 bonds found in alkanes and diamond, provide nanotubes with their unique strength. Within this chemical bonding, moreover, they align themselves into ropes held together by Van der Waals forces and can merge together under high pressure, trading some sp^2 bonds to sp^3 and producing very strong wires of nanometric lateral dimension.

The physical properties of nanotubes make them potentially useful in nanometric-scale electronic and mechanical applications. Currently, their physical properties are still being discovered and disputed. Nanotubes have a very broad range of electronic, thermal, and structural properties that change depending on the different kinds of nanotube (defined by their diameter, length, and chirality, or twist). In terms of mechanical strength, CNTs are considered to have a high tensile strength and elastic modulus, making them strong, stiff materials. Considerable research efforts are underway to modify CNT properties such as chirality, purity, length, and surface properties for binding with various materials [26–31]. Based on their characteristic properties, CNTs are identified as an ideal material for energy storage, conductive adhesive, inks, and grease, reinforcing fillers, catalyst supports, and many other advanced applications.

2.2.2.1 Single-Walled Carbon Nanotubes

SWNTs are nanometer-diameter cylinders consisting of a single GNS wrapped around to form a tube. Since their discovery in the early 1990s, there has been intense activity exploring the electrical properties of these systems and their potential applications in electronics. Experiments and theories have shown that these tubes can be either metals or semiconductors, and their electrical properties can rival, or even exceed, the best metals or semiconductors known. The first studies on metallic tubes were done in 1997 [32, 33] and the first on semiconducting tubes in 1998 [33]. In the intervening five years, a large number of groups have constructed and measured nanotube devices, and most major universities and industrial laboratories now have at least one group studying their properties.

Most SWNTs have a diameter of close to 1 nm, with a tube length that can be many millions of times longer. The structure of the SWNT can be theorized by covering a one-atom-thick layer of graphite, called graphene, into a seamless cylinder. The way the GNS is wrapped is represented by a pair of indices (n,m). The integers n and m denote the number of unit vectors along two directions in the honeycomb crystal lattice of graphene. If $m = 0$, the nanotubes are called zigzag nanotubes, and if $n = m$, the nanotubes are called armchair nanotubes. Otherwise, they are called chiral. The diameter of an ideal nanotube can be calculated from its (n,m) indices as follows:

$$d = \frac{a}{\pi}\sqrt{n^2 + nm + m^2} = 78.3\sqrt{\left((n+m)^2 - nm\right)}\,pm$$

where $a = 0.246nm$.

SWNTs are an important variety of CNT because most of their properties change significantly with the (n,m) values, and this dependence is non-monotonic (refer Kataura plot). In particular, their band gap can vary from 0 to about 2 eV and their electrical conductivity can show metallic or semiconducting behavior. SWNTs are likely candidates for miniaturizing electronics. The most basic building block of these systems is the electric wire, and SWNTs with diameters of an order of a nanometer can be excellent conductors [34, 35]. One useful application of SWNTs is in the development of the first intermolecular field-effect transistors (FET). The first intermolecular logic gate using SWNT FETs was made in 2001 [36]. A logic gate requires both a p-FET and an n-FET. Because SWNTs are p-FETs when exposed to oxygen and n-FETs otherwise, it is possible to protect half of an SWNT from oxygen exposure while exposing the other half to oxygen. This results in an SWNT that acts as a NOT logic gate with both p- and n-FETs within the same molecule. SWNTs are dropping precipitously in price, from around $1500 per gram as of 2000 to retail prices of around $50 per gram of as-produced 40%–60% by weight SWNTs as of March 2010.

SWNTs show remarkable electrical properties. Their band structure is quite unusual and is called a *zero-band gap semiconductor* since it is metallic in some directions and semiconducting in the others. In an SWNT, the momentum of the electrons moving around the circumference of the tube can be quantified. This quantization results in tubes that are either 1D metals or semiconductors, depending on how the allowed momentum states compare to the preferred directions for conduction. The tube acts as a 1D metal with a Fermi velocity $v_f = 8 \times 10^5$ m s^{-1} comparable to typical metals. If the axis is chosen differently, the tube acts as 1D semiconductor, with a band gap between the filled hole states and the empty electron states. The band gap is predicted to be $E_g = 0.9$ eV d^{-1}, where d is the diameter of the tube. Nanotubes can, therefore, be either metals or semiconductors, depending on how the tube is rolled up. This remarkable theoretical prediction has been verified using a number of measurement techniques.

2.2.2.2 Multi-Walled Carbon Nanotubes

MWNTs consist of multiple rolled layers (concentric tubes) of graphene. There are two models that can be used to describe the structures of MWNTs: the "Russian Doll" model and the "Parchment" model. In the Russian Doll model, sheets of graphite are arranged in concentric cylinders, each SWNT within a larger SWNT. In the Parchment model, a single sheet of graphite is rolled in around itself, resembling a scroll of parchment or a rolled newspaper. The interlayer distance in MWNTs is close to the distance between graphene layers in graphite, approximately 3.4 Å. The Russian Doll structure is observed more commonly. Its individual shells can be described as SWNTs, which can be metallic or semiconducting. Because of statistical probability and restrictions on the relative diameters of the individual tubes, one of the shells, and thus the whole MWNT, is usually a zero-gap metal.

Double-walled CNTs (DWNT) form a special class of nanotubes because their morphology and properties are similar to those of SWNTs but their resistance to chemicals is significantly improved. This is especially important when functionalization is required (this means grafting of chemical functions at the surface of the nanotubes) to add new properties to the CNT. In the case of SWNTs, covalent functionalization will break some C=C double bonds, leaving "holes" in the structure of the nanotube and thus modifying both its mechanical and electrical properties. In the case of DWNTs, only the outer wall is modified. DWNT synthesis on the gram scale was first proposed in 2003 [37] by the chemical vapor deposition (CVD) technique, from the selective reduction of oxide solutions in methane and hydrogen. The telescopic motion ability of inner shells Cumings [38] and their unique mechanical properties [39] will permit the use of MWNTs as the main movable arms in future nanomechanical devices. A retraction force occurs due to the telescopic motion caused by the Lennard-Jones interaction between shells, and its value is about 1.5 nN [40].

2.2.3 Graphene

The discovery of graphene [41] heralded a breakthrough in the scientific community. The 2D one-atom-thick planar sheet of the sp^2 hybridized carbon atoms [42, 43] possesses very unique properties compared to conventional materials [44]. It is the thinnest, most flexible, and strongest material ever discovered. The exceptionally high surface area, exceptional electrical conductivity, and excellent optical properties make it a very important engineering material. Since 2005, when the exceptional electronic properties of GNS were reported, there has been a gold rush among materials scientists in terms of exploiting the full range of properties of GNS. Table 2.1 summarizes some of the important properties of GNS [45, 46].

Carbon-based materials such as activated carbon, CNTs, and buckyballs have been widely utilized as active electrodes for energy devices. Structural polymorphism, excellent chemical stability, wide potential windows, inert electrochemistry, high surface area, and electrocatalytic activities for a variety of redox reactions, coupled with its low cost, make activated carbon the prime anode material for commercial LIBs and supercapacitors. But with the superior properties of graphene reported so far, it is envisioned to be the material that will revolutionize the battery industry.

TABLE 2.1   Important Inherent Properties of
Graphene Nanosheets

Electron mobility	2×10^5 cm^2 V^{-1} s^{-1}
Resistivity	10^{-6} Ω
Thermal conductivity	5×10^3 W m^{-1} K^{-1}
Optical transparency	97.7%
Specific surface area	2630 m^2 g^{-1}
Breaking strength	42 N m^{-1}
Elastic modulus	0.25 TPa
Ultimate tensile strength	130 GPa
Young's modulus	15×10^7 psi (1 TPa)
Spring constant	1–5 N m^{-1}
Fracture toughness	~4 MPa√m

Apart from electrochemical properties, high surface area and electronic conductivity are essential characteristics in an electrode material for energy storage or energy generation applications [47]. Graphene has a theoretical surface area of 2630 m^2 g^{-1}, which is two times higher than CNTs (1315 m^2 g^{-1}) and 250 times higher than graphite (~10 m^2 g^{-1}). The electronic conductivity of graphene is stable over a wide range of temperatures, and such stability is essential for reliability within a plethora of applications. The electronic conductivity of graphene is reported to be ~64 mS cm^{-1}, which is about 60 times higher than SWNTs, and its electron mobility is 200,000 cm^2 V^{-1} s^{-1} at room temperature, which is 200 times higher than silicon [47]. Interestingly, graphene is distinguished from its counterparts by its unusual band structure, in which the quasiparticles are formally identical to the massless Dirac Fermions [47]. The extreme electronic quality of graphene is further indicated by its Fermi velocity. Graphene displays the half-integer quantum Hall effect, with the effective speed of light as its Fermi velocity, $V_F \approx 10^6$ m s^{-1}, which can be observed in graphene even at room temperature. The quality of graphene clearly reveals itself with a pronounced ambipolar electric field effect; charge carriers can be tuned continuously between electrons and holes where electron mobility remains high even at high concentrations in both electrically and chemically doped devices, which translates to ballistic transport on the sub-micrometer scale. Due to its unique properties, GNS can carry a supercurrent [48], and it is clear that its theoretical electron transfer rates are superior when contrasted with other carbon allotropes. Furthermore, the fast charge carrier properties of graphene were found not only to be continuous but also to exhibit high crystal quality. Compared to its counterparts, in graphene crystallites, the charge carriers can travel thousands of inter-atomic distances without scattering [43], demonstrating its exceptional electronic qualities. Compared to all other potential materials, graphene has exhibited the fastest electron mobilities *theoretically*, meaning that in many electrochemical applications, graphene-based electrodes can outperform a majority of materials.

2.3 INHERENT ELECTROCHEMISTRY OF GRAPHENE NANOMATERIALS

Graphene is a 2D single atomic planar sheet of sp^2 bonded carbon atoms. They are the key derivative of carbon and originate from a large family of fullerene nanomaterials, wherein it is the essential "building block" for many of the allotropic dimensionalities that have significant and widespread use as electrode materials. In addition to existing in its planar state, graphene can be "wrapped" into 0D spherical buckyballs, "rolled" into 1D CNTs, or stacked into 3D graphite where stacks generally consist of more than ten graphene sheets or graphene nanoplatelets, believed to have between two and ten graphene sheets stacked upon one another. A single GNS is essentially equivalent to SWNTs, and graphene nanoplates are equivalent to MWNTs in that multiple layers of graphene exist. Most of the interesting properties of graphene derive from its truly 2D densely packaged honeycomb lattice structure. Within the lattice, the carbon atoms are sp^2 hybridized, and hence each carbon is connected via σ bonds to its three neighbors, having a bond length of 1.42 Å and an angle between the bonds of 120°. These σ bonds are also responsible for the robustness of the lattice.

GNS becomes spontaneously oxidized by contact with air and/or through the employed fabrication process, producing either chemically or electrochemically reduced graphene oxide (rGO). The properties of graphene strongly depend on their structure and on the number and/or position of the oxygen-containing groups. The conductivity of graphene oxide (GO) is poor, and thus reduction of GO is mandatory for most of the electrochemical applications [49]. GO sheets exhibit a significant reduction between −0.60 V (vs. Ag/AgCl reference electrode) and −0.87 V, and the reduction process significantly depends on the pH of the medium. The conductivity of the electrochemically reduced graphene oxide increases by eight orders of magnitude versus graphene oxide [50]. It should be noted that the mechanism of electrochemical reduction is not yet fully understood. However, in general, the capacitance of graphene nanomaterials greatly depends on the number of 2D layers stacked together to form the electrode in electrochemical devices. The number of stacking layers of multilayer GNS can be engineered by selective post-treatment and the number of 2D layers can be determined by specific surface area. The high superior electrochemical properties of engineered graphene-based electrodes are attributed to the dependence of the space charge layer capacitance of graphene on the number of layers, where the two factors of screening length and stacking thickness have primary influence.

As highlighted above, graphene holds inimitable properties that are superior in comparison to other carbon allotropes of various dimensions and to any other electrode material for that matter, thus *theoretically* suggesting that graphene is an ideal electrode material that could yield significant benefits in many electrochemical applications.

2.4 MECHANISM OF LITHIUM INSERTION IN CARBON NANOMATERIALS

Graphene and CNTs are the allotropic forms of graphite, offering LIBs higher energy density as well as alleviating the risk of pulverization with graphite due to their unique morphology, high tensile strength, high conductivity, and relative inertness to chemical degradation [51–53]. SWNTs and graphene are expected to exhibit reversible capacities somewhere around 300–600 mAh g^{-1} [54–57], effectively surpassing graphite-based anodes. The capacities of CNT/graphene-based electrodes can further enhance by preparing hybrid nanocomposite electrodes with other high-capacity electrode materials [13, 58, 59].

As a promising candidate for advanced LIBs, graphene/CNTs have been well studied to understand the mechanism of intercalation/deintercalation, adsorption, and diffusion of Li^+-ions into the graphene/CNTs during charge/discharge cycles, both theoretically and experimentally. Two different mechanisms, intercalation and alloy formation are proposed [60] for the anode materials. Intercalation compounds include allotropes of carbon (graphene and CNTs), TiO_2 compounds, $T-Nb_2O_5$, and so on, while alloying is usually observed in metals. Intercalation materials exhibit fast Li^+-ion reaction kinetics and structural integrity (owing to low volume expansions), which is desirable for the high-rate capability and long cycling stability of LIBs [61]. Hence, this section restricts the discussion on the intercalation/deintercalation process of Li^+-ions at different sites on carbonaceous nanostructures applicable to graphene/CNTs. So far, the intercalation/deintercalation process of Li^+-ions in CNTs are more complicated and thus Li^+-ion intercalation/deintercalation are explained more specifically for CNTs.

Structurally, CNTs are flawless 1D cylinders composed of one or more rolled-up GNS, with an aspect ratio of >1000. The interior as well as exterior surfaces of CNTs are electrochemically active for Li intercalation/absorption, justifying the multitude of research activities focused on their role as performance boosters of both the electrodes in LIBs [62, 63].

Among the studies performed to investigate the Li^+-ion intercalation mechanism in CNTs through electrochemical intercalation of Li into raw end-closed CNTs, including theoretical works

[64–70], Yang et al. [64] proposed a surface mechanism by which the naked surface of CNTs and carbon nanoparticles could store Li species. Since the investigations were on end-caped pristine CNTs, the Li^+-ions adsorbed can only be localized at the surface of CNTs.

The studies applied with first-principles methods carried out to elucidate the nature of Li intercalation in CNT ropes of both zigzag SWNTs and armchair SWNTs show charge transfer between Li and carbon and a small deformation of CNT structures due to intercalation. The study found that both the interior and interstitial spaces on the nanotubes were susceptible to Li intercalation and achieved a high Li^+-ion density [65]. *Ab initio* quantum chemical calculations designed to clarify which sites (the side-wall or cap region of CNTs) are more favorable for Li^+-ion intercalation revealed the dependence of the barrier for Li insertion on the size of the carbon ring. Insertion, being easier for larger rings, preferred two positions: the interior of the tube near the wall, and outside the tube [71]. The adsorption of a single Li^+-ion on the inside of the SWNTs is favored compared to the outside, since the Li adsorption energy of one Li atom on the inside is −0.98 eV, while that on the outside is −0.86 eV (Senani et al. [66]). It was also found that after the Li^+-ion attachment, the charge is transferred from Li^+-ions to CNTs, and the bonds between Li atoms and CNTs have ionic properties. The amount of charge transfer is greatly dependent on the radii of curvature of CNTs. The multiple attachments of Li^+-ions on the inside of the CNT model showed that four Li insertions into one-layer SWNTs is the most stable.

The interstitial spaces of CNT bundles resulting from Van der Waals forces are expected to carry a higher capacity for Li^+-ion intercalation and, consequently, higher reversible Li^+-ion storage capacity, which is demonstrated with a chemically etched CNT bundle [67]. These findings are again supported by studies on the intercalation and diffusion of Li^+-ions in a CNT bundle by *ab initio* molecular dynamics simulations [68]. The studies found that

Li$^+$-ions can penetrate quickly into CNTs and into the interstitial spaces between neighboring CNTs. Interestingly, a new mechanism for Li intercalation and storage was observed in that Li$^+$-ions could intercalate between two or more neighboring CNTs. The Li$^+$-ions, however, are located in the cloud of three neighboring CNTs that have very strong adsorption potential, which make it difficult to remove Li$^+$-ions from the bundles of nanotubes. This corresponded favorably to the irreversible Li$^+$-ion storage capacity reported in etched CNT bundles [67].

The diffusion of Li$^+$-ions in the CNT tubes can exist in two directions: the radial direction and the axial direction [69]. Studies on the interaction and diffusion of Li$^+$-ions in (5,5) armchair CNTs using density-functional theory showed that, for a single Li$^+$-ion moving in the radial direction, the most favorable positions for Li$^+$-ions are along a straight line passing through the center of a six-membered carbon ring on one side of the wall, and the midpoint of a C–C bond on the opposite side. The studies on interaction between Li$^+$-ions found that the interaction energy per Li atom as a function of Li-Li separation clearly indicates an oscillatory character that is especially obvious at larger separations. This can be attributed to two separate effects: incomplete screening from the tube and the Li-tube interaction that depends on the position of Li with reference to the hexagonal carbon framework. Compared to the independent diffusion of individual atoms, the oscillatory ion–ion repulsion favors the concerted diffusion of many Li atoms along the axis [70]. This agrees well with previous reports, in which Li$^+$-ions display high mobility along the tube axis with an energy barrier of <47 meV, while the diffusion barrier along the radial direction is as high as 380 meV.

The intercalation capacity in CNTs is intimately linked to the morphology of the nanotube under investigation and is no longer limited to LiC$_6$ [72–74]. Defects in CNTs may either be naturally occurring or artificially induced; either way, the defect affects the morphologies of the nanotubes, thereby influencing their Li$^+$-ion storage capacities [56, 72, 75, 76]. Defects in CNTs are introduced by removing carbon atoms from the regular hexagonal rings of side walls and that have been quantified. The removal of carbon atoms from CNTs makes a "hole" in the side walls, which becomes larger with the simultaneous removal of more carbon atoms. It is reported that Li^+-ions rarely diffuse inside defect-free $(n=6)$, $(n=7)$ (one C–atom removed) and $(n=8)$ (two C-atoms removed) nanotubes, while it can easily diffuse inside $(n=9)$ (three C-atoms removed) nanotubes. Upon entering the CNTs, the Li^+-ions are free to move in the interior of the CNTs and can also accumulate on the exterior wall [77–79]. For open-ended CNTs, the Li^+-ions also can enter the CNTs through the ends of the CNTs. It has been reported that the Li^+-ion storage capacity increases from LiC_6 for close-ended CNTs to LiC_3 in open-ended CNTs [72]. However, to enable a reasonable rate of ion exchange, the tube length of open-ended CNTs must be short [80]. Length plays quite an important role in the diffusion coefficient of CNTs, due to the fact that after insertion, Li^+-ions undergo a 1D random walk inside the nanotubes. However, if the length of the nanotube is long, the intercalated Li^+-ions are unable to exit from the nanotube, thereby decreasing the effective diffusion [80–82]. Electron density analysis has demonstrated that complete charge transfer takes place between Li^+-ions and CNTs as soon as Li intercalates the nanotube [65]. Thus, the fundamental factor dominating the Li^+-ion kinetics in CNTs is the competency of Li^+-ions to reach any of its open ends or interstitial channel [80]. While there is generally an increase in the reversible capacity of CNTs with the introduction of defects, irreversible capacity is also amplified. This means that some of the diffused Li^+-ions are consumed by the anode during the first cycle and are never restored at the cathode; therefore, no networking is done by these "lost" ions.

As previously mentioned, depending on their chirality, CNTs can be either metallic or semiconducting. Metallic CNTs are found to exhibit ~400% more Li insertion capacity than semiconducting ones. This type of behavior has been ascribed to (a) differences in the electrical conductivity of the two types of CNTs, and (b) disparities in Li^+-ion absorption potentials for both the nanotubes [57]. From the above discussion, it seems that CNTs can be used to improve the capacities of LIBs.

2.5 SUMMARY

The last few decades represent a golden period in the science of carbon. As a result of unconditional scientific efforts and extensive research into on all possible carbon allotropes, carbon is one of the materials of the millennium. This chapter discusses different carbon allotropes and summarizes the properties as well as applications of CNTs and graphene for the development of high-performance LIBs. Compared to other carbon allotropes, CNTs and graphene have gained much attention in the field of electrochemical storage devices and especially for LIBs, due to their unique structure and excellent electrical, thermal, mechanical, chemical, and electrochemical properties. The chapter also discusses the inherent electrochemical properties and the mechanism of Li^+-ion storage of CNTs and graphene for LIBs.

REFERENCES

1. Whittingham et al. *Science* 192 (1976) 1126.
2. Whittingham et al. *Chem. Rev.* 104 (2004) 4271.
3. Besenhard et al. *J. Electroanal. Chem.* 53 (1974) 329.
4. Besenhard et al. *Carbon* 14 (1976) 111–115.
5. Besenhard et al. *J. Electroanal. Chem. Interfacial Electrochem.* 68 (1976) 1.
6. Schöllhorn et al. *Mater. Res. Bull.* 11 (1976) 83.
7. Besenhard et al. *J. Power Sources* 1 (1976) 267.
8. Eichinger et al. *J. Electroanal. Chem. Interfacial Electrochem.* 72 (1976) 1.
9. Basu et al. *Mater. Sci. Eng.* 38 (1979) 275.
10. IMLB, 1982, C.L.U.P. Ed. Milan, Abstract #23.
11. Yazami et al. *J. Power Sources* 9 (1983) 365.
12. Howard et al. *J. Power Sources* 165 (2007) 887.
13. Reddy et al. *Nano Lett.* 9 (2009) 1002.
14. Susantyoko et al. *RSC Adv.* 8 (2018) 16566.
15. Wang et al. *ACS Nano* 4 (2010) 2233.

16. Osawa et al. *MRS Bull.* 19 (1994) 33.
17. Cataldo. *Carbon* 40 (2002) 157.
18. Kudryavtsev et al. *Russ. Chem. Bull.* 42 (1993) 399.
19. Lagow et al. *Science* 267 (1995) 362.
20. Heinmann et al. *Carbon* 35 (1997) 1654.
21. Iijima et al. *Nature* 354 (1991) 56.
22. Lin et al. *J. Mater. Chem.* 14 (2004) 527.
23. Moniruzzaman et al. *Macromolecules* 39 (2006) 5194.
24. Wang et al. *Nano Lett.* 9 (2009) 3137.
25. Gullapalli et al. *Chem. Eng. Prog.* 107 (2011) 28.
26. Thostenson et al. *Compos. Sci. Technol.* 61 (2001) 1899.
27. Bethune et al. *Nature* 366 (1993) 123.
28. Ren et al. *Science* 282 (1998) 1105.
29. Journet et al. *Nature* 388 (1997) 756.
30. Rinzler et al. *Appl. Phys. A Mater. Sci. Process.* 67 (1998) 29.
31. Nikolaev et al. *Chem. Phys. Lett.* 313 (1999) 91.
32. Bockrath et al. *Science* 275 (1997)1922.
33. Tans et al. *Nature* 386 (1997) 474.
34. Mintmire et al. *Phys. Rev. Lett.* 68 (1992) 631.
35. Dekker. *Phys. Today* 52 (1999) 22.
36. Martel et al. *Phys. Rev. Lett.* 87 (2001) 256805.
37. Flahaut et al. *Chem. Commun.* 12 (2003) 1442.
38. Zettl et al. *Science* 289 (2000) 602.
39. Treacy et al. *Nature* 381 (1996) 678.
40. Zavalniuk et al. *Low Temp. Phys.* 37 (2001) 337.
41. Novoselov et al. *Science* 306 (2004) 666.
42. Meyer et al. *Nature* 446 (2007) 60.
43. Geim et al. *Nat. Mater.* 6 (2007) 183.
44. González et al. *Carbon* 50 (2012) 828.
45. Mao et al. *RSC Adv.* 2 (2012) 2643.
46. Reina et al. *Nano Res.* 2 (2009) 509.
47. Brownson et al. *Analyst* 135 (2010) 2768.
48. Sato et al. *Fujitsu Sci. Tech. J.* 46 (2010) 103.
49. Dreyer et al. *Chem. Soc. Rev.* 39 (2010) 228.
50. Zhou et al. *Chem. Eur. J.* 15 (2009) 6116.
51. Chan et al. *Nat. Nanotechnol.* 3 (2008) 31–35.
52. Cui et al. *Nano Lett.* 9 (2009) 3370–3374.
53. Yu et al. *Science* 287 (2000) 637–640.
54. H.S. Oktaviano et al. *J. Mater. Chem.* 22 (2012) 25167–25173.
55. Li et al. *Electrochem. Commun.* 12 (2010) 592–595.
56. Yang et al. *Electrochim. Acta* 53 (2008) 2238–2244.

57. Kawasaki et al. *Mater. Lett.* 62 (2008) 2917–2920.
58. Eom et al. *J. Electrochem. Soc.* 153 (2006) A1678–A1684.
59. Zhang et al. *Adv. Mater.* 21 (2009) 2299–2304.
60. Casas et al. *J. Power Sources* 208 (2012) 74.
61. Hayner et al. *Annu. Rev. Chem. Biomol. Eng.* 3 (2012) 445.
62. Liu et al. *Appl. Phys. Lett.* 84 (2004) 2649.
63. Zhao et al. *Phys. Rev. B* 71 (2005) 165413.
64. Yang et al. *Mater. Chem. Phys.* 71 (2001) 7.
65. Zhao et al. *Phys. Rev. Lett.* 85 (2000) 1706.
66. Senami et al. *AIP Adv.* 1 (2011) 042106.
67. Shimoda et al. *Phys. Rev. Lett.* 88 (2002) 015502.
68. Song et al. *Energy Environ. Sci.* 4 (2011) 1379.
69. Khantha et al. *Phys. Rev. B* 78 (2008) 115430.
70. Zhao et al. *Phys. Lett. A* 340 (2005) 434.
71. Kar et al. *J. Phys. Chem. A* 105 (2001) 10397.
72. Shimoda et al. *Phys. Rev. Lett.* 88 (2001) 015502.
73. Chen et al. *Carbon* 41 (2003) 959–966.
74. Yang et al. *Electrochim. Acta* 52 (2007) 5286.
75. Gao et al. *Chem. Phys. Lett.* 327 (2000) 69.
76. Mi et al. *J. Electroanal. Chem.* 562 (2004) 217.
77. Nishidate et al. *Phys. Rev. B* 71 (2005) 245418.
78. Garau et al. *Chem. Phys. Lett.* 374 (2003) 548.
79. Garau et al. *Chem. Phys.* 297 (2004) 85.
80. Meunier et al. *Phys. Rev. Lett.* 88 (2002) 075506.
81. Wang et al. *J. Power Sources* 186 (2009) 194.
82. Sehrawat et al. *Mater. Sci. Eng. B* 213 (2016) 12.

CHAPTER 3

Carbon Nanotube and Its Composites as Anodes for Lithium Ion Batteries

3.1 INTRODUCTION

Materials that have the capability to store Li^+-ions are widely used as negative electrodes (anodes) in lithium ion batteries (LIBs). There are three basic requirements for anode materials: (i) The potential of Li insertion/extraction in the anode versus Li must be as low as possible, (ii) the amount of Li that can be accommodated by the anode material should be as high as possible to achieve a high specific capacity, and (iii) the anode host should endure repeated Li insertion/extraction without any structural damage to obtain long cycling stability. Traditionally, graphitic carbon has been used for anodes in commercial LIBs since 1991, when Sony commercialized LIBs for the first time. Graphite is a 2D planar material formed by stacking hexagonally bonded thin sheets of carbon material called *graphene nanosheet* (GNS), which are held together by Van der Waals forces. During charging, Li^+-ions generate from the cathode and intercalate between graphite layers in a configuration of one Li^+-ions for every six carbon atoms, leading to the formation of LiC_6 [1, 2]. Typically, graphite anodes are electrochemically stable, environmentally friendly, and economically viable; however, they have a poor specific capacity of only 372 mAh g^{-1}, which is much lower than the specific capacity of Li metal anodes. Even though Li metal anodes have a very high capacity of 3860 mAh g^{-1}, they are commercially not that attractive due to their high reactivity with electrolyte, dendrite formation, and related safety issues [3]. The search for high-capacity anodes has led to the discovery of a series of materials such as Al, Sn, Si, Bi, and Sb that can alloy with Li to form high-capacity anode materials for LIBs [4–8]. However, these materials show poor cyclability and mechanical stability due to the mechanical cracking caused by huge volume expansion/contraction during the alloying and dealloying reactions with Li^+-ions [8]. There are many approaches such as the application of pressure on cells, using elastomeric binders or forming a composite with conductive materials [9–12]. Carbon nanotube (CNT), or graphene, has garnered much attention due to its nanosize, high conductivity, large electrochemically accessible surface area, and so on. This chapter focuses on the application of CNT not only as a conducting-cum-mechanically integrating filler in composite anodes but also its use as a binder-free and flexible anode, as well as the development of CNT/graphene hybrid anodes for advanced LIBs. Among the different composite anodes, the chapter is limited to only Sn-based anodes.

3.2 FREE-STANDING CARBON NANOTUBE ANODE

Many efforts have been made to use CNTs as a replacement for graphite in Li^+-ion cells, and both single-walled CNTs (SWNTs) and multi-walled CNTs (MWNTs) have been studied as anode materials in LIBs. SWNTs can have a reversible capacity from 300 to 600 mAh g^{-1} [13–19], which is significantly higher than the capacity of graphite (372 mAh g^{-1}). The reversible capacity of SWNTs can be further improved up to 1000 mAh g^{-1} by chemical and mechanical treatments [20–23]. The use of hybrid composite materials with CNT as a critical component represents a practical way to enhance the charge capacity and to reduce the irreversible capacity of LIBs [24–27]. However, the presence of the binder still leads to a lowering of the specific capacity and efficiency of the anode due to its insulating nature. In this scenario, the cutting-edge technology is the concept of the ability to maintain bifunctionality, where CNTs can act as both the active material as well as the current collector, involves using free-standing CNT paper as an anode in LIBs. These paper-like materials are lightweight, flexible, binder-free, and thermally stable, making them suitable for even high-temperature applications. Transferable, conductive CNT films open up a new avenue for high-performance flexible batteries due to their high electrical conductivity, flexibility, and low density [28]. Currently, many preparation methods are used to manufacture CNT films or papers, such as the pressure filtration method [28], the layer-by-layer method [29], the rod coating method [30], and chemical vapor deposition (CVD) [31].

Fabrication of both SWNT [32] and MWNT [33] free-standing flexible papers having stable thermal and mechanical properties has been reported. SWNTs easily aggregate to form bundles having a quasi 2D hexagonal lattice due to weak Van der Waals interactions, thereby functioning as an efficient ion storage material. The ordered CNT bundles consist not only of the huge amount of inner space of the tube but also the huge intertubular nanospace in the bundle, which act as the sites for Li intercalation. It was reported that the reversible capacity of the chemically treated SWNTs per unit weight is about three times higher than graphite, the commercial anode material in LIBs [34]. However, the chirality of the CNTs plays a crucial role in Li storage, because the conductivity greatly depends on the chirality. Kaiser et al. [35] studied the conduction mechanism in CNT, and Kawasaki et al. [18] studied the electrochemical Li$^+$-ion storage properties of SWNT bundles with different electronic conductivity, one with metallic conductivity and other semiconducting [18, 35]. SWNTs were prepared by a pressure filtration technique, whereby dispersions of SWNTs are eluted through an inert and porous support material, typically like Teflon, and peel off after drying as free-standing flexible papers [32]. These papers show a tensile strength of ~80 to 100 MPa and Young's modulus of 5–10 GPa, which suggests that a large compressive as well as tensile force can be applied prior to plastic deformation. Their good mechanical integrity makes these CNT papers attractive for the fabrication of flexible LIBs as they can be shaped into any required form and size and can be punched easily using conventional cutters or punchers. These SWNT papers show a conductivity of about 5×10^5 S m^{-1} [35]; hence, when used as an anode, the electronic transport can be similar to metallic conduction and exhibit high specific capacity, which varies with the thickness of the electrode due to the difference in mass loading [36]. The initial reported capacities of flexible free-standing SWNTs have been found to be between 400 and 460 mAh g^{-1}, which have been further improved by shortening the length of the SWNTs and by introducing sidewall defects. This has resulted in a specific capacity of close to 1000 mAh g^{-1} [37–44] being achieved, due to the metallic conductivity, high surface area, and bundling effect of CNTs.

SWNTs produced by the HiPco method were wrapped in nickel mesh after purification and were used as the anode in LIBs. It is reported that as the conductivity increases, the Li$^+$-ion storage capability or specific capacity increases. The metallic SWNTs delivered a discharge capacity of 445 mAh g^{-1}, which is ~3.4 times greater than that of semiconducting SWNT at a current density of 100 mA g^{-1}. Raman studies showed that the semiconducting SWNT samples have 100% semiconducting nanotubes, while the metallic SWNTs have a mixture of metallic and semiconducting SWNTs; hence, the reversible capacity of the metallic SWNTs cannot be determined by the charge/discharge profiles of the metallic SWNT. The reversible capacity of the 100% metallic SWNT is calculated to be 641 mAh g^{-1}, which is five times higher than that of semiconducting SWNT, suggesting that the reversible Li$^+$-ion storage capacity of a metallic SWNT is about five times greater than that of a semiconducting SWNT [18]. The significantly higher capacity shown by metallic SWNTs is attributed to their higher Li$^+$-ion adsorption potentials [45] and the kinematic effect caused by the difference in electrotonic conductivity.

Flexible free-standing anode for LIBs using bare MWNTs [33, 46–48] and alumina-coated MWNTs [49] were studied by Gao et al. [46], who compared the electrochemical properties of ordered versus defective MWNTs, while the study by Yoon et al. [47] focused on nitrogen-doped (N-) fiber shaped MWNTs. These studies showed that defective MWNTs have better Li storage capacity and electrochemical stability than ordered ones due to the fact that defects in MWNT films allow easier insertion/deinsertion of Li$^+$-ions as well as lower charge transfer resistance. The defective MWNT flexible anode showed a reversible capacity of 452 mAh g^{-1}, which is ~20% higher than ordered crystalline MWNT anodes (375 mAh g^{-1}) at a current density of 30 mA g^{-1}, and they showed a similar percentage capacity fade of ~77% after 500 cycles at 3000 mA g^{-1} current density [46]. The first cycle discharge capacity of the N-doped MWNT anode was 3600 mAh g^{-1}, which is ~2.8 times greater than pristine MWNT at 0.5 C; however, the discharge capacity falls sharply (to about 800 mAh g^{-1} for N-doped MWNT) during the second discharge cycle and starts to stabilize after ten cycles. After ten cycles, the cell delivers a discharge capacity of 446 and 165 mAh g^{-1} respectively for N-doped and pristine MWNTs and maintains stability through 50 cycles [47].

When transferable conductive MWNT films prepared using the CVD method were used as a free-standing anode after transferring the film on to graphene-coated polyethylene terephthalate (PET), an initial charge/discharge capacity of 397/899 mAh g^{-1} at 0.1 C was delivered. The specific capacity of these free-standing films is ~381% (248 mAh g^{-1}) higher than a recently reported CNT anode on a copper foil (~65 mAh g^{-1}). At 1 C, a specific capacity was produced that was 248 mAh g^{-1} higher than that of the recently reported CNTs on a Cu anode (~65 mAh g^{-1}) [50], free-standing CNTs [28, 50], and vertically aligned CNTs/conducting polymer composites [51]. The rate capability studies show that once the C-rate returns to the initial 0.1 C from 3 C, the MWNT anodes recover their average specific capacity of 374 mAh g^{-1}, suggesting that the electrodes maintain their structural integrity and physicochemical properties after experiencing higher current densities [33].

Alumina-coated CNTs showed extremely high reversible specific capacity, cycling stability, and rate capability [49]. The first cycle reversible capacity was 3036 mAh g^{-1} at 38 mA g^{-1} current density, which is a much higher capacity than the reversible capacity delivered by free-standing MWNT anodes [33, 46–48]; for example, a defective MWNT flexible anode showed a reversible capacity of only 452 mAh g^{-1} even at a low current density of 30 mA g^{-1} [46] and a stable capacity of ~1100 mAh g^{-1} after six cycles [49]. Compared to previous reports [48], the reversible specific capacity of aluminum coated MWNT has been shown to be considerably higher than the anode without aluminum coating prepared by the same CVD method. At a current density of 38 mA g^{-1}, the aluminum coated MWNT anode shows an initial reversible specific capacity ~540 mAh g^{-1} [49] higher than the MWNT anode without aluminum coating. The enhanced electrochemical properties and exceptional "zero capacity degradation" during long cycle operation showed by the aluminum coated MWNT is due to the excellent conductivity, structural integrity, and Li^+-ion intercalation ability offered by the core of the structure. The good bonding between the substrate and MWNTs and the bonding between the MWNTs in the nanotube forest creates a good interface that leads to minimal contact resistance or ohmic contact, and the MWNT structure does not show any expansion/contraction problems during lithiation/delithiation and hence poses no risk of pulverization. A very high surface area of MWNT is available for lithiation, and easy ion transport through the highly porous

CNT forest structure leads to excellent cycling stability and rate capability. The aluminum oxide coating provides additional thermomechanical stability, along with further enhancement of the reversible capacity and safety of LIBs [49].

3.3 CARBON NANOTUBE OF DIFFERENT MORPHOLOGIES AS ANODES FOR LIBS

The electrochemical performance of CNTs is greatly influenced by their morphology; that is, the degree of defects, the length of the buckytubes, and the diameter of the CNTs play a crucial role in the performance of CNT-based anode materials. The morphological difference can result from different synthesis methods, processing techniques, or physical/chemical modifications. Commonly, there are two widely used methods to modify the morphologies of CNT: chemical etching and mechanical milling. The processing technique includes electrospinning [52, 53], twisting of CNT bundles, dry/wet spinning [54], or the direct spinning of CNT fibers from CVD [55]. These methods result in structural changes and/or the formation of surface functional groups on CNTs. The structural changes include lateral defects on the surface of CNTs and/or shortening of the length of CNTs due to the breaking of the tube parallel to the radial axis, which can improve the Li storage capacity. Morphology features such as the alignment, open/closed end, bundle size, and diameter of the CNTs can be controlled by adjusting the synthesis parameters such as temperature, pressure, constituent components, and catalyst. The following section discusses the effect of morphology on the electrochemical properties and Li storage capability of CNTs.

3.3.1 Effects of Defects on Electrochemical Properties of CNTs

The presence of defects subtle alerts the properties of all materials such as chemical reactivity, mechanical strength, optical absorption, and electronic transport. The extent of property changes greatly depends on the concentration of defects. For the certain applications, the defects do not greatly affect the performance of the device; for example, an electronic circuit performs perfectly well in the presence of defects. However, a defect might trap or scatter charge carriers. During research into their applications in LIBs, defective CNTs were reported to be more effective for the adsorption and diffusion of Li^+-ions [43, 56–59]. Anodes based on pristine CNTs generally suffer from low practical capacities and high irreversible charge loss due to the formation of a solid electrolyte interface (SEI) layer resulting from the chemical reaction between anode and electrolyte and/or other side reactions of anode/electrolyte during Li^+-ion intercalation/deintercalation [60]. These drawbacks of pristine CNTs can be overcome to some extent by introducing defects on the CNTs. These defects include holes in the side wall, the opening of the end cap, fragmentations located at the edges, and the formation of surface-bound oxygen and surface adsorbed fulvic acid. The presence of holes on the side walls of the CNTs or the removal of the end cap of CNTs ensures better intercalation and diffusion of Li^+-ions inside the CNTs; this can open up more active sites for the Li adsorption/diffusion, thereby increasing/enhancing the Li storage capacity. As the size of the defective ring increases, the energy barrier for the Li^+-ion diffusion decreases; this phenomenon has been experimentally proven by investigating the energetics of Li^+-ion adsorption on defective SWNTs [56]. The results showed that the diffusivity of Li^+-ions into pristine CNTs is difficult, while it is easier for defective CNTs. Compared to $n = 7$ and $n = 8$ defected CNTs, the Li diffusion is difficult as compared with $n = 9$ defected CNTs. The *ab initio* calculations also showed that the energy barrier for the diffusion through the nine-membered ring is more difficult than ten-membered rings. The *ab initio* calculations showed that the energy barrier for the diffusion through nine-membered rings is 9.69 kcal mol^{-1}, which is higher than that of $n \geq 10$ [61], and these results agree well with the studies carried out by Ishidate et al. [56].

There have been many studies carried out to understand the effect of defects on Li storage capacity prepared using the CVD method [46, 57, 59] morphological modification by gas-phase oxidation [60], and drilled CNTs by solid state reactions [62]. Gao et al. [46] compared the Li$^+$-ion storage capacity of defective and ordered crystalline CNTs, while Eom et al. [43, 57, 58] studied the kinetics of Li intercalation/deintercalation into MWNTs prepared using the CVD method. The defective CNT is prepared by varying the injection rate of liquid precursor [46] or by chemical etching [14, 43, 57, 58]. Both studies showed that defective MWNTs have more Li storage capacity. The electrochemical measurements in LIBs show that defective CNTs have a reversible capacity of 452 mAh g^{-1} at a current density of 30 mA g^{-1}, which is considerably higher than that of ordered crystalline (375 mAh g^{-1}) MWNTs under the same current density and testing conditions [46]. It is well known that the Li$^+$-ion storage capacity of CNTs increases with the number of defects, which is validated with experimental data [57]. The studies showed that the number of defects and the surface area, as well as the reversible and irreversible capacity, increases with etching time. The reversible capacity is improved from 351 (Li$_{0.9}$C$_6$) to 681 mAh g^{-1} (Li$_{1.8}$C$_6$), and the irreversible capacity increases from 1012 (Li$_{2.7}$C$_6$) to 1229 mAh g^{-1} (Li$_{3.3}$C$_6$) with etching time. The structural and chemical modifications of the etched MWNTs facilitate the insertion of Li$^+$-ions into the buckytubes via open ends and enhances the reversible capacity; however, the large surface area of the etched MWNTs induces large irreversible capacity, that is, the extraction of Li$^+$-ion from the etched MWNTs was greatly hindered. Compared with pristine MWNTs, the first discharge derivative curves showed a peak shift from 0.8 to 0.9 V for the etched MWNTs, indicating that Li$^+$-ions were inserted into the inner core of the etched MWNTs through the open ends; however, the extraction of Li$^+$-ions from the inner core was greatly hindered due to the strong tendency of MWNTs to accumulate or trap Li$^+$-ions. The improved electrochemical properties of etched MWNT samples are presumably due to the formation of the SEI on the large surface area by the chemical etching, while the increase in irreversible capacity is attributed to the increase of disordered or amorphous carbon at the expense of etching time [43, 57, 58]. Mi et al. [14] also investigated the effect of chemical etching on the structure and electrochemical properties of MWNTs. A large number of defects and pores were introduced on the walls and innertubes of MWNTs by the etching process, which significantly improves the electrochemical properties. As anode materials in LIBs, it is found that a considerably increased specific capacity is mainly in the voltage range above 1.0 V. The significantly higher reversible capacity and cycling stability of etched CNTs is related to Li doping into such regions as disordered graphitic structures, defects, microcavities, and edges of graphitic sheets or layers [14].

Klink et al. [60] oxidized MWNTs by either liquid or gas phase nitric acid in order to study the correlation between the amount, the type of surface oxygen group, the extent of exfoliation, and Li$^+$-ion storage capacity. The study showed that the method and duration of oxidation time have a pronounced effect on the morphology of MWNTs. The average length of CNT is found to be ≤ 1 μm for liquid-phase oxidation (at 85°C for 24 h treatment), while gas-phase oxidized CNT showed an average tube length of ≥ 5 μm (200°C 72 h treatment). A significant weight loss (25% loss for liquid-phase and 15% for gas-phase oxidation for 24 h) is reported, which is probably due to the filtration loss of small MWNT fragments or solubilized fluvic acid. The lower weight loss observed with gas-phase oxidized CNT may also be an indication of the introduction of more surface functional groups [60]. The electrochemical studies suggest that liquid-phase oxidized CNTs suffer from a substantially high amount of total irreversible charge loss that decreases with oxidation time from 283 mAh g^{-1} (1.5 h treatment) to 232 mAh g^{-1} (24 h treatment), which may come at cost of a shorter tube length. Both MWNTs show a very similar irreversible charge loss (~140 mAh g^{-1}) in the range of 3–5 V, suggesting a very similar active surface area and amount of SEI. However, gas-phase oxidized CNTs show a significantly lower irreversible charge loss of below 0.5 V, suggesting that the SEI is directly related to the third type of irreversible reaction, exfoliation, and occurs below 0.5 V versus Li^+/Li. When a solvated Li^+-ion penetrates the SEI layer without losing its solvation shell, the solvent molecules also get reduced and the decomposition products lever apart the graphene layers, causing exfoliation [63]. For gas-phase oxidized CNTs, the exfoliation charge loss was only 60 mAh g^{-1} (24 h) and 33 mAh g^{-1} (74 h) compared to 146 mAh g^{-1} (1.5 h) and 107 mAh g^{-1} (24 h) for liquid-phase oxidized CNTs. Liquid-phase oxidized CNTs (1.5 h) and gas-phase oxidized CNTs (24 h) are very similar in terms of morphology, the total amount of oxygen, and the surface oxygen concentration. The deintercalation capacity for liquid-phase oxidized CNT decreases from 174 mAh g^{-1} (1.5 h) to 149 mAh g^{-1} (24 h), which is related to the reduced graphitic carbon content (from 93 to 83 wt.%), as oxidation transforms a considerable amount of sp^2-hybridized carbon to oxygen-bound carbon. A significantly considerable increase in deintercalation capacity at a potential higher than 1.9 V is observed, rising from 87 mAh g^{-1} (1.5 h) to 231 mAh g^{-1} (24 h); however, the deintercalation capacity of liquid-phase oxidized CNTs is lower than that of gas-phase oxidized CNTs, even though liquid-phase oxidized CNTs have a higher carbon content. This suggests that it will be more difficult for Li^+-ions to intercalate into graphene layers. Therefore, it is supposed that the gas-phase oxidized CNTs are more suitable for use as anode material in LIBs since they show a lower specific initial charge loss as well as easy processing and less mechanical degradation [60].

Apart from chemical etching with strong acid, a novel chemical method called *nanodrilling* is reported to create multiple holes in the walls of MWNTs using CoOx as an oxidation catalyst. The electrochemical characterization of a nanodrilled MWNT-based anode showed a high specific discharge capacity of 625 mAh g^{-1} at a current density of 25 mAh g^{-1} after 20 cycles, while the capacities of the pristine and purified MWNTs were 267 and 421 mAh g^{-1}, respectively. The relatively higher capacity of nanodrilled MWNT results from increased Li storage sites, easy insertion/extraction of

Li$^+$-ions in/out of the inner core of the tube, and the capacity for the largest quantity of Li to be extracted from the graphite intercalation compound.

In addition, nanodrilled MWNTs show the largest quantity of functional groups of 5.4×10^{-3} mol g^{-1}, which suggests that the drilling method also introduces functional groups on the edges. In the first cycle, the columbic efficiency of the drilled MWNTs is 40% which is 2%–6% higher than the pristine/purified MWNTs. The rest of the capacity on the first cycle is related to irreversible processes, mainly the result of the reductive decomposition of the electrolyte and the formation of the SEI layer. Such a low initial columbic efficiency has been found in various nanocarbon materials because the large surface area will result in a large volume of SEI. The study also showed that the size of the holes drilled on the basal planes of MWNTs has a positive effect on the reversible capacity [62].

Except for CNTs with holes on the basal planes, open-ended CNTs can also show improved Li$^+$-ion storage capability. The electrochemical properties of open-ended MWNTs versus closed-ended MWNTs have been studied [42, 57, 59]. The caps of CNTs are opened by chemical etching of SWNTs [42] and MWNTs [57] by placing MWNTs between silica wool plugs in the silica tube and heating them under a flow of air at 700°C [59]. In all these studies, both SWNT [42] and MWNT [57, 59] open-end samples show a significantly improved reversible capacity than closed-ended CNTs when used as anode material in LIBs. A discharge capacity of 450 mAh g^{-1} is delivered by open-ended CNTs, which is larger than that of closed-ended CNTs. This result has been concurrent with other studies [40, 64] The higher capacity is considered to be the consequence of the Li$^+$-ion intercalation into the inner core of the CNTs through the open ends, which is not possible with closed-ended CNTs. Maurin et al. [64] and Gao et al. [40] expected that the open caps would give a much higher reversible capacity; however, the study by Yang et al. shows that it is difficult to deintercalate the Li$^+$-ion from the inner core of the

CNTs [59] due to the strong tendency of MWNT to accumulate charges as a condenser [65]. Hence, the electrostatic attraction of charged species seems to be a great hindrance to the extraction of Li$^+$-ions from the inner core of the CNTs, which is evident from the cycling stability of open-ended MWNTs [59]. The open-ended CNTs showed poorer cycling performance than the closed-ended CNTs due to the smaller Warburg prefactor and ionic conductivity value of closed-ended CNTs [42, 57, 59].

In summary, the open-ended or defective CNTs are more preferable than closed-ended or ordered CNTs as anode materials in LIBs due to their higher specific reversible capacity, electrochemical active surface area, shorter diffusion path or/and a lower diffusion barrier, lower Warburg prefactor, and impedance.

3.3.2 Effects of Length on Electrochemical Properties of CNTs

The length of the CNT is one of the important influential factors determining the morphological, mechanical, electronic, and electrochemical properties of the nanotubes. CNT synthesized by CVD, electric arc-discharge, and laser vaporization are usually several to tens of micrometers long with a high aspect ratio, tangled texture, and closed ends [66], all of which were unfavorable for Li$^+$-ion insertion/extraction when CNTs were used as anode materials for LIBs. The length of the CNT greatly influences the Li$^+$-ion intercalation and diffusion, electronic conductivity, and Li storage capability. Electrochemical studies of short and long CNTs as anode materials in LIBs showed that shorter CNTs have better charge/discharge properties and cycling stability as these CNTs facilitate easier intercalation/deintercalation of Li$^+$-ions. When inserted in the CNT, Li$^+$-ions undergo a 1D random walk inside the CNT, and effective diffusion/deintercalation will decrease if the tube is too long, that is, the Li$^+$-ions are able to enter the tube but seldom exit. Commonly used methods such as chemical etching, ball-milling, nanodrilling, electron irradiation, ion irradiation, solid-state cutting for modifying the morphologies of CNTs not only create defects in CNT structures but also cut CNTs into shorter lengths [20, 43, 46, 57, 67]. However, CNTs produced by chemical etching, ball milling, and nanodrilling contain a large amount of surface functional groups, which leads to substantial voltage hysteresis [43, 57, 58]. The large voltage hysteresis of CNTs may also relate to the kinetics of Li diffusion into the inner tubules of MWNTs, and this can be reduced by cutting the nanotubes into shorter segments [20].

Comparative studies of the electrochemical properties of short and long CNTs prepared by sonication [68], co-pyrolysis [69, 70], common CVD [66], solid-state cutting [71], and catalytic [72] methods have been reported. Shimoda et al. [68] found that the SWNTs could be cut into shorter CNTs with open ends and that the reversible capacity of CNTs increased from LiC_6 in closed-ended CNTs to LiC_3 after cutting due to the Li insertion into the CNT core through the open ends and sidewall defects [68]. Yang et al. made a comparative study on the electrochemical properties of short and long CNTs [69, 70]. The short CNTs have a length of 150–400 nm while that of the long CNTs is more than several micrometers. The electrochemical studies showed that the short CNTs have a reversible capacity of 266 and 170 mAh g^{-1} at current densities of 0.2 and 0.8 mA cm^{-2} respectively, which is about two times higher than the value obtained with the long CNTs [70]. Similar electrochemical performance is reported with short CNTs with an average tube length of 200 nm, prepared by solid-state cutting of conventional micrometer long entangled CNTs [69]. The solid-state cutting method involves depositing Ni onto CNTs, followed by oxidation of Ni at 400°C. The NiO particles produced react with CNTs at a high temperature of 900°C, resulted in the cutting of CNTs into a shorter length. As an anode material in LIBs, the short CNTs showed significantly higher reversible capacity; however, the irreversible capacity is slightly increased or even decreased as the length of CNT decreases. The cycling studies showed that the specific capacities became stable after 30 cycles for both electrodes. Again, the retention of specific capacities after 50 cycles was higher for short MWNTs compared to the long ones. Other electrochemical properties such as electronic conductivity, Warburg prefactor, impedance, and ionic conductivity result in better rate capability and cycling stability in short CNTs [71]. In subsequent work, Wang et al. [72] reported for the first time on the use of Fe compound (FeS) as a catalyst for directly controlling the length of CNTs. The directly grown short CNTs, having a tube length of 200–500 nm and 20–30 mm in diameter, were tested as anodes in LIBs and were compared with the electrochemical properties of short CNTs prepared by solid-state cutting and conventional long CNTs. The cell studies showed that the reversible capacity of the long CNTs is 188 mAh g^{-1}, while the reversible capacity of directly grown short CNTs and the cut CNTs dramatically increased to 502 and 577 mAh g^{-1} respectively at a current density of 25 mA g^{-1}, which clearly suggests that the short CNTs have a higher Li extraction capacity than the long CNTs. However, the irreversible capacities for both short and cut CNTs increased, which may be associated with the open ends, shortened tube length, and increased surface area. The initial coulombic efficiency for the long CNTs (31%) is considerably lower than the short CNTs (40%–49%). The cycling studies showed that after 20 cycles, the charge capacity of both short and long CNTs becomes stable. This stable capacity at a current density of 25 mA g^{-1} is about 230 mAh g^{-1} for the short CNTs and 142 mAh g^{-1} for the long CNTs, suggesting a better cyclic performance for the short CNTs. Both the short and the cut CNTs show similar electrochemical properties that are much better than those for conventional long CNTs. The reason for this may be the fact that Li^+-ion insertion/extraction into/from short CNTs is facilitated by the shortened length of the tubes. Moreover, short CNTs have much more edges (open ends) than long CNTs per unit length, making it much easier for Li^+-ion to intercalate/deintercalate into/from the graphitic sheets, resulting in higher reversible and irreversible capacities.

3.3.3 Effect of Diameter on the Electrochemical Properties of CNTs

The diameter is the parameter of CNTs that can most significantly influence the Li+-ion insertion/extraction into/from the CNTs. Immediately after Li+-ions intercalate into the CNTs, they take a 1D random walk inside the CNTs and fall in love with carbon atoms, which enhances the charge transfer between CNTs and Li+-ions [73]. This interaction between the CNT and Li+-ions is related to the curvature (diameter) of the tubes and will result in different Li storage capacities in CNTs [74]. The diameter of the MWNTs is also considered to influence the exfoliation of graphene sheets that are rolled up to form the multilayered structure of MWNT, and it will consequently influence the Li storage capacity [75]. Liu et al. [76] studied Li+-ion absorption energy in CNTs and local chemical bonding between Li-C and C-C using the first-principles molecular orbital method. The study showed that the Li absorption energy and binding energy of CNTs are greatly dependent on the CNT diameter. The first-principles total energy calculations of Li absorption into CNTs with various diameters showed that as the diameter of the nanotube increases, the Li outside-absorption energy decreases whereas the Li inside-absorption energy increases. For CNTs with a smaller diameter, the Li absorption energy is larger for outside than inside absorption; however, when the nanotube diameter increases more than 0.824 nm, the Li outside-absorption energy tends to reach a value similar to the Li inside-absorption energy. This suggests that the energetic tendencies of the Li outside and inside absorption would be very similar for larger nanotubes. When the tube diameter is 0.556 nm, the Li inside absorption may not occur because of the negative absorption energy. For the binding energy, the values for both the pristine and Li-absorbed nanotubes increase with the tube diameter. The study also showed that distinctive chemical interactions take place between Li and C atoms and between the C-atoms due to the curvature of carbon sheet in the Li-absorbed CNTs, which is somewhat different from those in Li-absorbed graphite or graphene with flat carbon sheets. The present results for the pristine nanotubes agree well with the results calculated from the continuum elastic model [77]. According to this model, the binding energy of a pristine nanotube is proportional to the square of the nanotube diameter, and it increases with the tube diameter in a similar way to the Li outside-absorbed nanotube. As the diameter increases, the binding energy of pristine nanotubes would converge to the value of graphite, and in the case of Li-absorbed nanotubes, to the value of LiC_6 [76]. Compared to the Liu et al. [76] study more directly, the research performed by Zhao et al. [78] used first principle calculation to explore the potential energy profile of Li confined inside SWNTs and the subsequent condensation process. The study showed a clear relationship between the Li/C ratio and the tube diameter. It was found that with an increase of the tube diameter, the intercalated Li atoms tend to form a multi-shell structure at the equilibrium state, which will improve the Li^+-ion storage capacity. The calculations based on first principles show that Li trapped inside SWNTs has high mobility along or around axial directions, whereas the radial motion is constrained. The energy barrier for the mobility of Li^+-ions around the tube axis is ⁵47 meV, whereas the diffusion barrier along the radial direction can be as high as 380 meV. This leads to the condensation of Li^+-ions on the walls of CNTs when they are placed randomly into SWNTs, resulting in nanowires with single or multi-shelled morphologies, depending on the diameter of the SWNT. The charge transfer from Li nanowires to SWNTs is significant, indicating a stronger coupling between them. The theoretical investigations into SWNTs using *ab initio* and molecular interaction potential with polarization methodologies showed that the interaction potential at the central region is dependent on the diameter of the nanotube, and that CNTs with a diameter of 4.68 Å have greater interaction energy with Li^+-ions, making them the best candidate to store Li^+-ions and consequently improve the performance of nanotube-based LIBs [74]. Recently, Zhang et al. [79] studied the electrochemical properties of MWNTs with the aim of understanding the effect of tube diameters of CNTs on the electrochemical properties of MWNTs. They compared the morphological and electrochemical properties of MWNTs with different diameters (10–100 nm) as anode materials in LIBs. The influence of diameter on electrochemical properties of MWNTs was investigated using charge/discharge cycling tests and electrochemical impedance spectroscopy. The results revealed that MWNTs with a tube diameter of 40–60 nm showed the first discharge and charge capacity of 318 and 162 mAh g^{-1} at a current density of 50 mA g^{-1} and stable cycling performance. After the 50th cycle, these electrodes showed the highest specific capacity of 187 mAh g^{-1} at a current density of 50 mA g^{-1}, with a columbic efficiency of ~102%. In addition, they showed least SEI film resistance and a charge transfer resistance of 2.6 Ω and 42.2 Ω, respectively. It was observed that the charge transfer and SEI film resistance decreases with the tube diameter ranging from 10 to 60 nm, which suggests that they can preserve the highest conductivity. Hence, as an anode in LIBs, electron transport is greatly enhanced during the Li intercalation/deintercalation process, which results in significant advantages in the electrochemical performance of MWNTs as an anode in LIBs.

3.4 SUMMARY

The constantly increasing demand for electrical storage for portable electronic devices and sustainable transportation is spurring the development of thinner, lighter, more flexible, and more efficient batteries with high energy density, cycling stability, and rate capability. The design flexibility in materials and fabrication methods provided by the current state-of-the-art LIB technology accelerated the development of novel anode materials and electrode fabrication methods. This chapter discussed recent developments in anode research and the new opportunities opened up by the use of carbon materials, especially SWNTs and MWNTs, as anode materials in LIBs. As CNTs have unique electrochemical properties and Li$^+$-ion storage mechanisms, the chapter focuses on the effect of the morphology, structure, and dimension of CNTs on the electrochemical performance of LIBs. The long tubular structure and high mechanical strength of CNTs allow the fabrication of free-standing electrodes (without binder/current collector) as an active Li^+-ion storage material or physical support for ultra-high-capacity material. The use of extraordinary electronic conductivity and the Li^+-ion storage capability of CNTs have been demonstrated both theoretically and experimentally, and thus they are logically considered ideal anode materials for high energy LIBs. The electrochemical properties of CNTs, especially their Li^+-ion storage capacity and rate capability, can be further enhanced by opening the nanotube ends and separating chiral fractions; in a nutshell, the need to develop high-performance LIBs will fuel research on nanoscale materials like CNT or graphene in order to capitalize on their unique size-dependent properties for the significant improvement of cutting-edge technologies.

REFERENCES

1. Dahn et al. *Phys. Rev. B* 44 (1991) 9170.
2. Satoh et al. *Solid State Ionics* 80 (1995) 291.
3. Reddy et al. 2002. *Handbook of Batteries*, ed. Linden, D., Reddy, T. B., 3rd ed., 34.1–34.62.
4. Besenhard et al. *J. Power Sources* 68 (1997) 87.
5. Winter et al. *Electrochim. Acta* 45 (1999) 31.
6. Winter et al. *Adv. Mater.* 10 (1998) 725.
7. Obrovac et al. *Electrochem. Solid State Lett.* 7 (2004) A93.
8. NuLi et al. *Mater. Lett.* 62 (2008) 2092.
9. Liu et al. *Electrochem. Solid State Lett.* 8 (2005) A100.
10. Liu et al. *Electrochem. Solid State Lett.* 8 (2005) A599.
11. Guo et al. *J. Electrochem. Soc.* 152 (2005) A2211.
12. Wu et al. *J. Power Sources* 114 (2003) 228.
13. Yan et al. *Electrochim. Acta* 53 (2008) 6351.
14. Mi et al. *J. Electroanal. Chem.* 562 (2004) 217.
15. Nishidate et al. *Phys. Rev. B* 71 (2005) 245418.
16. Meunier et al. *Phys. Rev. Lett.* 88 (2002) 075506.
17. Kumar et al. *Electrochem. Commun.* 6 (2004) 520.
18. Kawasaki et al. *Mater. Lett.* 62 (2008) 2917.

19. Morris et al. *J. Power Sources* 138 (2004) 277.
20. Gao et al. *Chem. Phys. Lett.* 327 (2000) 69.
21. Shimoda et al. *Phys. Rev. Lett.* 88 (2002) 015502.
22. Yang et al. *Mater. Lett.* 50 (2001) 108.
23. Maurin et al. *Solid State Ionics* 136–137 (2000) 1295.
24. Reddy et al. *Nano Lett.* 9 (2009) 1002.
25. Zhang et al. *Int. J. Nanomanuf.* 2 (2008) 4.
26. Eom et al. *J. Electrochem. Soc.* 153 (2006) A1678.
27. Zhang et al. *Adv. Mater.* 21 (2009) 2299.
28. Chew et al. *Carbon* 47 (2009) 2976.
29. Lee et al. *Energy Environ. Sci.* 4 (2011) 1972.
30. Hu et al. *Proc. Natl. Acad. Sci. U S A* 106 (2009) 21490.
31. Zhong et al. *Adv. Mater.* 22 (2010) 692.
32. Gupta et al. *J. Solid State Electrochem.* 16 (2012) 1585.
33. Kang et al. *Nanotechnology* 27 (2016) 105402.
34. Komiyama et al. *Tanso* 216 (2005) 25 (in Japanese).
35. Kaiser et al. *Physica E* 40 (2008) 2311.
36. Jin et al. *Electrochem. Commun.* 10 (2008) 1537.
37. Claye et al. *J. Electrochem. Soc.* 147 (2000) 2845.
38. Mukhopadhyay et al. *Physica B* 323 (2002) 130.
39. Ng et al. *Electrochim. Acta* 51 (2005) 23.
40. Gao et al. *Chem. Phys. Lett.* 307 (1999) 153.
41. Gao et al. *AIP Conf. Proc.* 590 (2001) 95.
42. Shimoda et al. *Phys. Rev. Lett.* 88 (2001) 015502.
43. Eom et al. *J. Mater. Res.* 23 (2008) 2458.
44. Jeong et al. *Langmuir* 17 (2001) 8281.
45. Nishidate et al. *Phys. Rev. B* 71 (2005) 245418.
46. Gao et al. *Vacuum* 112 (2015) 1.
47. Yoon et al. *J. Power Sources* 279 (2015) 495.
48. Lahiri et al. *ACS Nano* 4 (2010) 3440.
49. Lahiri et al. *J. Mater. Chem.* 21 (2011) 13621.
50. Li et al. *J. Mater. Chem.* 22 (2012) 18847.
51. Chen et al. *Chem. Mater.* 19 (2007) 3595.
52. Zhang et al. *Angew. Chem. Int. Ed.* 53 (2014) 14564.
53. Chen et al. *Energy Environ. Sci.* 5 (2012) 7898.
54. Jestin S., Poulin P. 2014. *Nanotube Superfiber Materials Changing Engineering Design*, ed. Schulz, M. J., Shanov, V. N., Yin, Z. Elsevier, 167–209.
55. Li et al. *Science* 304 (2004) 276.
56. Ishidate et al. *Phys. Rev. B* 71 (2005) 245418.
57. Eom et al. *Carbon* 42 (2004) 2589.

58. Eom et al. *J. Power Sources* 157 (2006) 507.
59. Yang et al. *Solid State Ionics* 143 (2001) 173.
60. Klink et al. *Electrochem. Commun.* 15 (2012) 10.
61. Garau et al. *Chem. Phys. Lett.* 374 (2003) 548.
62. Oktaviano et al. *J. Mater. Chem.* 22 (2012) 25167.
63. Spahr et al. *J. Electrochem. Soc.* 151 (2004) A1383.
64. Maurin et al. *Chem. Phys. Lett.* 312 (1999) 14.
65. Frackowiak et al. *Carbon* 37 (1999) 61.
66. Ren et al. *Science* 282 (1998) 1105.
67. Pierard et al. *Chem. Phys. Lett.* 335 (2001) 1.
68. Shimoda et al. *Physica B* 323 (2002) 133.
69. Yang et al. *Electrochem. Commun.* 8 (2006) 137.
70. Yang et al. *Electrochim. Acta* 53 (2008) 2238.
71. Wang et al. *Adv. Funct. Mater.* 17 (2007) 3613.
72. Wang et al. *J. Power Sources* 186 (2009) 194.
73. Senami et al. *AIP Adv.* 1 (2011) 042106.
74. Garau et al. *Chem. Phys.* 297 (2004) 85.
75. Zhang et al. *J. Electrochem. Commun.* 13 (2011) 125.
76. Liu et al. *Comput. Mater. Sci.* 30 (2004) 50.
77. Robertson et al. *Phys. Rev. B* 45 (1992) 12592.
78. Zhao et al. *Phys. Lett. A* 340 (2005) 434.
79. Zhang et al. *Appl. Surf. Sci.* 258 (2012) 4729.

CHAPTER 4

Carbon Nanotube and Its Composites as Cathodes for Lithium Ion Batteries

4.1 INTRODUCTION

The development of the positive electrode is one of the key factors in achieving optimum performance in lithium ion batteries (LIBs). Among different factors such as electronic, ionic, and thermal conductivities, thermomechanical stability is significantly important in the design of LIBs with superior energy storage capacity, prolonged life, and high levels of safety for high-performance applications such as electric vehicles, portable electronic devices, and power grids. Carbon nanotubes (CNTs) are widely studied as an anode as well as cathode material in LIBs due to their unique structural and physical features, which include excellent electrical, mechanical, and chemical properties. Their large mesoporosity and high electrical conductivity make them suitable candidates for cathode materials in electrochemical energy storage devices. The incorporation of CNTs can improve the electrochemical properties of cathode materials without compromising their physical properties. Layer-structured $LiMnO_2$, $LiNiO_2$, $LiCoO_2$, spinel $Li(Ni_xMn_y)O_2$, $Li(Ni_xCo_yMn_z)O_2$, $Li(Ni_xMn_y)O_4$, $LiMn_2O_4$, ordered olivine Li-transition metal phosphates $LiMPO_4$ (M = Fe, Mn, Ni, or Co), and elemental sulfur are widely studied cathode materials in LIBs [1–12]. $LiCoO_2$, $LiMn_2O_4$, and $LiFePO_4$ have been used as cathode materials in commercial LIBs, and hence this chapter will focus on the electrochemical properties of composite cathodes made by combining these active materials with multi-walled CNTs (MWNTs). It is well known that all these cathode materials share the notorious problem of poor electronic conductivity. The electronic conductivities of $LiCoO_2$, $LiMn_2O_4$, and $LiFePO_4$ are as low as 10^{-3}, 10^{-4}, and 10^{-9}, respectively, leading to the incomplete utilization of the active materials and severe polarization. Different methods have been adopted to solve this problem, including the modification of the active material by lattice doping, surface coating (e.g. carbon coating), and the addition of various conducting agents such as conducting polymer and different carbon allotropes. Among these, the incorporation of conductive additives is the most widely used strategy in the industrial production of LIBs. However, the commonly used conducting agents usually possesses poor crystallinity (acetylene black, Super P, or small surface area graphite) and conducting polymers, and they have limited efficiency in terms of their ability to form a continuous and highly conductive network through the composite electrode. As a result, the amount of these electrochemically inactive conductive agents is high as 10 wt.% for low-rate LIBs and even 20 wt.% for high-rate LIBs, which seriously reduces the energy densities of electrodes. To address this problem and reduce the amount of conducting additives, various types of carbon nanomaterials such as carbon nanofibers (CNFs), CNTs, and graphene have been employed to take advantage of their higher aspect ratio, mechanical integrity, electrochemical activity, and superiority in forming long-range conducting pathways. In many previous studies of CNT composite electrodes, CNTs are incorporated with active particles by either mechanical mixing [13, 14] or chemical bonding [15, 16]. The following section discusses the effect of CNTs, types, and functionalization on the electrochemical properties of popular cathode materials in LIBs.

4.2 CATHODES BASED ON TRANSITION METAL OXIDES/CNTs

4.2.1 LiCoO$_2$/CNT Composite Cathode

LiCoO$_2$ generally offers a higher capacity (260 mAh g^{-1}) compared to other metal oxides (LiMn$_2$O$_4$ and LiFePO$_4$); however, its poor electronic conductivity causes serious polarization and poor practical capacity. Efforts have been made to improve the specific capacity and cycling stability of LCoO$_2$ by making a composite cathode with carbon black (Super P), conducting polymers, carbon coating, and so on; however, these approaches show only nominal improvement in the electrochemical properties due to the limited surface area, poor electrical conductivity, and higher dosage. CNTs have been reported as an effective conductive additive to improve the electronic conductivity of LiCoO$_2$ and thus the electrochemical performance of the composite cathode. Both functionalized/doped or non-functionalized/undoped MWNTs as well as single-walled CNTs (SWNTs) have been used as conductive additives with LiCoO$_2$; however, MWNTs perform best. A comparison study of MWNT versus carbon black (acetylene black and Super P) as a conducting additive with LiCoO$_2$ was studied as a cathode material in LIBs. The electrochemical properties of the composite electrodes prepared by simple mixing showed that MWNTs are superior to carbon black in terms of both high C-rate performance and cycle life [17, 18]. LiCoO$_2$ cathode with 3 wt.% MWNTs or Super P delivers an initial discharge capacity of ~155 and 150 mAh g^{-1} respectively at a C-rate of 1 C after 40 cycles. The cathode with MWNT retains a discharge capacity of about 120 mAh g^{-1}, while the Super P cathode delivers a discharge capacity of ~100 mAh g^{-1} [17]. The initial discharge capacity of the LiCoO$_2$ cathode loaded with acetylene black or MWNTs at 0.5 C-rate was 134.4 or 139.2 mAh g^{-1} respectively [18]. The improved electrochemical performance of the LiCoO$_2$/MWNT cathode is due to the MWNT aggregates that form a conductive bridge between the particles of active material and maintain the intimate contact between the particles even when the composite expands on cycling; in contrast, similar but rigid bridges of carbon black are broken upon cycling [17, 18]. The electrochemical performance of the LiCoO$_2$ composite cathode incorporated with a mixture of MWNTs and acetylene black outperformed the cathode loaded with either MWNTs or acetylene black. In the LiCoO$_2$/MWNT/ acetylene black cathode, MWNTs forms a valid conducting network, while the acetylene black particles improve the contact surface area with LiCoO$_2$ particles [18]. The electrochemical performance of the LiCoO$_2$ composite cathode with various conductive additives (carbon black, carbon fibers, and MWNTs) shows a reduction in capacity retention ratios of 10% and 30% respectively for cathode material–containing carbon fibers and carbon black after 20 cycles, while the composite cathode containing MWNTs shows negligible capacity fades [19]. This was followed by the comparison of MWNTs with Super P [17] to evaluate the electrochemical properties of LiCoO$_2$ coated with 0.5 wt.% thin MWNTs (t-MWNTs) and hollow MWNTs (h-MWNTs) prepared by electrostatic heterocoagulation. The t-MWNTs showed superior electrochemical properties to h-MWNTs and showed ~65% (403 mAh cm^{-1}) discharge capacity retention with respect to an initial volumetric discharge capacity of 624 mAh cm^{-1} at a current density of 0.2 C [20]. The initial (0.2 C-rate) and after 40 cycles (1 C-rate) volumetric specific discharge capacity of the cathode with 3 wt.% MWNTs prepared by simple mixing was observed to be 546 and 310 mAh cm^{-1}, respectively [17], which is much less than the LiCoO$_2$/t-MWTS cathode [20]. The high volumetric capacity with a lower mount of t-MWNTs was attributed to the self-assembled nanotube network formed between active particles during coagulation and is well maintained with volume expansion during continuous charge/discharge cycling.

A comparison study was made on composite $LiCoO_2$ cathodes with SWNTs versus MWNTs [21] and MWNTs with different densities [22]. The electrochemical performance showed that the low-density [22] and smaller-diameter [21] CNTs exhibit excellent high-rate capabilities and cycle performance. The highest discharge capacity of 136 mAh g^{-1} was reported at 5 C-rate, and only 3% capacity fade was observed for composite $LiCoO_2$ cathodes having 8 wt.% MWNTs (low density) after 50 cycles at 1 C-rate [22].

A comparison study of $LiCoO_2$ composite cathode incorporated with SWNTs and MWNTs showed that the structure of the nanotube plays a crucial role in the electrochemical properties of the composite cathode. It was found that smaller-diameter nanotubes are better in terms of improving battery performance, especially at higher charge/discharge rates, due to their advantage in primary particle number in unit mass. The study showed that that at least 5 wt.% of MWNTs (dia. 10–30 nm) is a prerequisite if full use is to be made of $LiCoO_2$ active material. Due to their tendency toward bundle formation, SWNTs are not effective in $LiCoO_2$ composite cathodes compared with MWNTs [21].

Traditional electrodes in LIBs are often composed of binders, which are electrochemically inactive. Binders are commonly required to provide mechanical connections between active materials and conductive additives. The insulating nature of binders reduces the overall energy density by adding weight to the electrodes and leads to poor electron transfer on cycling. Thus, the elimination of binders will represent remarkable progress for high-performance LIBs. Efforts have been made to obtain binder-free CNT composite electrodes. For example, Fe_3O_4-SWNT electrodes were fabricated by dispersing SWNT and FeOOH nanorods in sodium dodecyl sulfonate solution followed by vacuum filtration and annealing processes [23], and binder-free CNT composite electrodes were prepared by loading functional particles onto CNT sheets drawn from CNT arrays by spraying, filtration, electrophoresis, evaporation, the sol-gel method, and so on [24, 25]. Binder-free $LiCoO_2$ super-aligned CNTs (SANTs) composite cathodes were prepared using an ultrasonication and co-deposition technique, which constructs a continuous 3D SANT framework embedding $LiCoO_2$ particles inside. The composite cathodes exhibited superior properties, such as high conductivity, great flexibility, and outstanding cycling stability (151.4 mAh g^{-1} at 0.1 C with cycle retention of 98.4% over 50 cycles) and rate capability (137.4 mAh g^{-1} at 2 C), to classical composite cathodes (active material, Super P, and binder) [26].

Very recently, a simple and feasible strategy of using cross-stacked SANT films as conductive layers to prepare sandwich-structured $LiCoO_2$ cathodes for high-performance LIBs has been reported. The sandwiched electrode structure, consisting of a repeating and alternating stack of $LiCoO_2$ layers and SANT films, ensures that each layer of active materials can adhere to the SANT conductive layers, resulting in the effective transfer of electrons throughout the electrodes regardless of the electrode thickness. Significant improvements in conductivity as well as cell performance are achieved with the introduction of three separate SANT conductive layers. The sandwich-structured $LiCoO_2$/Super P (2 wt.%)/SANT cathodes have an impressive rate capability (109.6 mAh g^{-1} at 10 C, which is ~1600% improvement compared with that without SANT films) due to the formation of a homogeneous and efficient conductive network in the electrodes owing to the super-aligned feature, the SANTs showing the best rate performances reported so far for commercial micro-sized $LiCoO_2$ particles. The easy fabrication procedure, compatible method for commercialization, low cost, and outstanding electrochemical performance of the sandwich-structured electrode demonstrate its great potential for the large-scale production of high-performance electrodes for LIBs [27].

4.2.2 Modified $LiCoO_2$/CNT Composite Cathode

Considering the safety and environmental issues related to $LiCoO_2$, partial replacement of Co was done by substituting

Al, Ni, Ga, Mg, Mn, or Ti, which overcomes the safety issues. Compared to other metals, Ni and Mn are widely used as a partial substitute for Co, and it is reported that new electrode designs have reduced weight. A composite cathode with partial replacement of Co, $Li(Ni_{1/3}Co_{1/3}Mn_{1/3})O_2$, $LiNi_{0.7}Co_{0.3}O_2$ [17, 28–31] $LiNi_{0.4}Mn_{0.4}Co_{0.2}O_2$ [32] and $Li_{1.17}Ni_{0.17}Co_{0.1}Mn_{0.56}O_2$ [33] and a complete replacement of Co, $LiNi_{0.5}Mn_{1.5}O_4$ [34] composite cathode incorporated with MWNTs have been reported. In all these cases, the incorporation of CNTs effectively improved the battery performance primarily due to the formation of efficient electron-conducting channels between the metal-oxide particles. For example, a 100 nm thick homogeneous MWNT coating on $Li_{1.17}Ni_{0.17}Co_{0.1}Mn_{0.56}O_2$, by applying shear stress without the breakdown of the crystal structure and morphology of the cathode, showed an electronic conductivity of 170 mS cm^{-1}, which is over 40 times higher than a pristine cathode with the same amount of conductive carbon powder (total carbon ratio is 94:3) due to the exceptionally good area between over-lithiated layered oxide (OLO) active material and carbon nanotubes. At a high current rate condition of 2450 mA g^{-1}, (10 C-rate), a specific capacity of 103 mAh g^{-1} is obtained even with only 3 wt.% carbon, including the coated MWNT. The cycling stability studies in full-cell configurations using graphite anode and 2 wt.% MWNT-coated cathode exhibited a higher discharge capacity, showing 150 mAh g^{-1} (divided by the mass of OLO) more than that containing the pristine sample at a constant current/constant voltage rate of 245 mAh g^{-1} between 2.5 and 4.6 V at 25°C and stable cycling even after 70 cycles [33]. The nanocomposite cathode with 95 wt.% $LiNi_{0.4}Mn_{0.4}Co_{0.2}O_2$ and 5 wt.% SWNTs, prepared by simple mixing, delivered a capacity of ~130 mAh g^{-1} at 5C and nearly 120 mAh g^{-1} at 10 C, both for over 500 cycles, and showed significantly higher capacity as compared with the pristine $LiNi_{0.4}Mn_{0.4}Co_{0.2}O_2$ at rates of 5–10 C. The addition of SWNTs to $LiNi_{0.4}Mn_{0.4}Co_{0.2}O_2$ resulted in exceptional improvements in both conductivity and surface stability and high performance at a high C-rate of 10 C (charge/discharge in 6 min) [32]. The free-standing, binder-free flexible cathode based on $LiNi_{0.5}Mn_{1.5}O_4$/MWNT delivered 80% of the discharge capacity at 1 C, and even when the charge/discharge current density increased to 20 C (capacity at 1 C = 140 mAh g^{-1}), there was no obvious capacity decay as observed after 100 cycles at a high C-rate of 10 C [34].

4.2.3 $LiMn_2O_4$ or $LiMnO_2$/CNT Composite Cathode

Other metal oxides such as $LiMn_2O_4$ [35–37] and $LiMnO_2$ [38, 39] nanocomposites with CNTs have also been reported to show improved electrochemical performance. $LiMn_2O_4$/MWNT nanocomposite cathodes prepared using facile sol-gel [35] and hydrothermal methods have been reported. The composite cathode is composed of active material, conducting carbon and binder (poly[vinylidene fluoride], PVdF) (70:20:10). Both the specific capacity and cyclic stability of the cell composed of a $LiMn_2O_4$-MWNT composite cathode and 1 M $LiPF_6$ in EC:DEC:EMC (1:1:1 vol.%) were observed to be superior to the pristine cathode. The $LiMn_2O_4$/MWNT composite cathode delivers a high specific capacity of 145.4 mAh g^{-1} and 0.1 C-rate, which is close to the theoretic capacity of $LiMn_2O_4$ (148 mAh g^{-1}), and it retained a discharge capacity of 114.8 mAh g^{-1} even if the charge/discharge current rate was pushed to 20 C. The cell also exhibits excellent cycling stability; when the cell is repeatedly cycled for 1000 cycles at 1 C-rate, the specific capacity is decreased from the initial value of 140.4 mAh g^{-1} to 98.7 mAh g^{-1}, which represents 70.3% capacity retention [36]. However, in a previous study, the $LiMn_2O_4$-MWNT cathode showed only 66.5 mAh g^{-1} and retained 99% of the initial discharge capacity after 20 cycles and a 4% capacity fade after 100 cycles, while $LiMn_2O_4$ showed a 9% loss of the initial capacity (54.7 mAh g^{-1}) after 20 cycles [35].

The high-performance flexible and binder-free $LiMn_2O_4$/CNT (MWNT) [37], coaxial $LiMnO_2$/CNT (MnO_2 shell and CNT core) [39], and $LiMnO_2$/CNT (aligned CNT, A-CNT) [38] composite cathodes were fabricated via an in-situ two-step hydrothermal process, vacuum infiltration in combination with chemical vapor deposition (CVD), and spontaneous reduction of $KMnO_4$ at 70°C in aqueous medium, respectively. The electrochemical performance of $LiMn_2O_4$/MWNT was compared with a pristine $LiMn_2O_4$ nanoparticle composite cathode (active material:conducting carbon:PVdF binder in 80:10:10) using 1 M $LiClO_4$ in PC. The $LiMn_2O_4$/MWNT flexible cathode shows an initial charge/discharge capacity of 126 mAh g^{-1}, while pristine $LiMn_2O_4$ showed only a capacity of 109 mAh g^{-1} at a current density of 22 mA g^{-1}. The $LiMn_2O_4$/MWNT cathode retained a discharge capacity of 50 mAh g^{-1} when the current density was pushed to a high value of 550 mA g^{-1} and showed stable cycling performance [37]. The $LiMnO_2$/CNT coaxial hybrid cathode showed very high initial charge/discharge capacity of 2170 mAh g^{-1} and a reversible capacity of ~500 mAh g^{-1} after 15 cycles at a constant current rate of 50 mA g^{-1}, which is much higher than MnO_2 nanotubes. The enhanced electrochemical performance is attributed to the improved conductivity achieved by providing highly conductive CNTs in the inner core of the MnO_2 nanotube shells, homogeneous electrochemical accessibility and high ionic conductivity by avoiding agglomerative binder, well-directed conductive paths due to perfect coaxial alignment, and a dual Li storage mechanism of insertion/deinsertion in the case of CNTs and formation and decomposition of Li_2O in the case of MnO_2 nanotubes. Also, the highly conductive CNT core acts as a buffer to alleviate the volume expansion caused by repeated ionic intercalation. The porous structure of the electrodes decreases the effective diffusion path and increases the surface area for the insertion/deinsertion of Li^+-ion; this method is quite difficult to scale up because of the alumina templates [39]. Followed by $LiMnO_2$/CNT coaxial hybrid cathode, as-synthesized $LiMnO_2$/A-CNT cathode prepared by Lou et al. [38] delivered a higher specific capacity equivalent to a theoretical capacity of MnO_2 (308 mAh g^{-1}) at a current density of 0.1 C and maintained an excellent rate capability (95 mAh g^{-1} at 20 C). However, the as-synthesized 3D nanostructured $LiMnO_2$/A-CNT cathode suffers from serious capacity fading due to the presence of trace water between the MnO_2 layers, and this problem is addressed by annealing the electrode for 12 h in argon. It is reported that the initial specific capacity decreases and cycling stability increases concurrently with the annealing temperature. After annealing at 200°C, the initial discharge capacity dropped from 308 to 245 mAh g^{-1} as a result of the partial collapse of MnO_2 layers, which reduces the accessibility of Li^+-ions to MnO_2. The annealed samples, however, maintained a specific capacity of 133 mAh g^{-1} after 100 cycles against 110 mAh g^{-1} for the unannealed sample at a current density of 1 C, which is attributed to the elimination of bound water molecules; to some extent, this leads to an increase in interaction between the manganese oxide and the A-CNTs [38].

4.2.4 Lithium Transition Metal Phosphates/CNT Composite Cathode

Li-transition metal phosphates (M-PO_4), where the transition metal can be Fe, Mn, Ni, or Co, have an olivine-type structure and a high theoretical specific capacity (~170 mAh g^{-1}) [40, 41]. Compared to other metal phosphates, $LiFePO_4$ is widely used in LIBs not only due to its high stability, compatibility, and environmental friendliness but also the fact that it is non-toxic and economically viable. However, complete utilization of $LiFePO_4$ to achieve full theoretical capacity is difficult owing to its poor electronic conductivity (~10^{-9} S cm^{-1}), which subsequently results in poor discharge capacity, cycle performance, rate capability, and Li^+-ion diffusion. Mixing with carbon black or conductive carbon coating are popular methods that have been adopted to improve the conductivity of $LiFePO_4$. However, cycling stability and rate capability are not significantly improved, and researchers have tried to address these problems by preparing composite $LiFePO_4$ cathodes using CNTs. Studies on the influence of the type, size, and amount of CNTs on the electrochemical performance of $LiFePO_4$ have been carried out for optimizing electrical resistivity, specific capacity, cycling stability, and rate capability. Various approaches such as microwave assisted synthesis [42], spray pyrolysis modified CVD techniques [43], the hydrothermal process [44], the facile in-situ sol-gel method [45–47], electrospinning [48], the microwave irradiating solvothermal method [49], solution-based impregnation [50], anodic electrode-polymerization [51, 52], and so on have be reported for preparing $LiFePO_4/CNT$ for preparing composite cathode. Both SWNT [53] and MWNT [51, 52] are used with $LiFePO_4$.

Carbon-coated $LiFePO_4/MWNT$ (C-$LiFePO_4/MWNT$) composite cathode having micron-sized $LiFePO_4$ (100–200 nm) [51, 52] and nanosized $LiFePO_4$ (40–90 nm) [54, 55] having 1–3 nm thick carbon coating using poly(acrylonitrile) (PAN) [51, 52] or sucrose [54, 55] as carbon source. The electrical conductivity of C-$LiFePO_4/MWNT$ is improved from 5 to 10 times with respect to pristine $LiFePO_4$. The C-$LiFePO_4/MWNT$ composite showed an electrical conductivity of 2.3×10^{-2} S cm^{-1} (micron-sized $LiFePO_4$) and 7.7×10^{-2} S cm^{-1} (nanosized $LiFePO_4$), which is higher than C-$LiFePO_4$ or pristine $LiFePO_4$, indicating that CNTs and amorphous carbon co-constructed conductive networks are more effective than single amorphous carbon in improving electronic conductivity owing to the more efficient electron transporting property of the graphitized CNT components. Due to the synergistic effect, the amorphous carbon coating improves the Li diffusion process, whereas CNTs construct an efficient conductive network to the optimized transport pathways for electrons. C-$LiFePO_4/CNT$ delivered a discharge capacity of ~170 mAh g^{-1} at a current density of 0.5 C and ~142 mAh g^{-1} at 20 C with good capacity retention [51, 52]. However, Wang et al. reported a capacity of 110 mAh g^{-1} at 20 C and 73 mAh g^{-1} at 60 C [54]. The C-$LiFePO_4/MWNT$ assures superior electrochemical properties in terms of an ultra-high-rate capability. About 59% capacity retention at the rate up to 120 C and a long cycling stability of 98.5% capacity retention over 500 cycles has also been reported [55]. For the practical applications of LIBs in electric vehicles, charging time is an important limiting factor. It is demonstrated that the C-LiFePO$_4$/CNT cathode took only 142 s (20 C) for charging 76% of the total charge capacity [54].

Qiao et al. reported a LiFePO$_4$/CNT nanocomposite cathode having ultra-long cycling stability prepared by CNTs coated with polyvinylpyrrolidone (PVP). The resulting composite cathodes containing 3 wt.% CNT showed a high discharge capacity of 123 mAh g^{-1}. Also, the composite cathode showed an extremely low capacity fade of 1.6% and 20% up to 1000 and 3400 cycles respectively at a current density of 10 C. The capacity loss in the study is 4–8 times smaller than previously reported LiFePO$_4$/CNTs and LiFePO$_4$/graphene composites. The excellent rate capability and ultra-low cycling stability is attributed to the synergistic effect of the large Li$^+$-ion diffusion coefficient achieved in LiFePO$_4$ nanoparticles and the highly conductive 3D network formed by the mono-dispersed CNTs, which is free from breaking and entangling effects [56]. LiFePO$_4$/CNT core-shell hybrid nanowires synthesized by aqueous solution-based mineralization under ambient conditions (20 wt.% CNT) [57], solution-based two-step method at 60°C [58], solid-state reaction at room temperature [59] and liquid deposition approach low temperature in an ice bath [60] are reported. A LiFePO$_4$/CNT core-shell delivered a discharge capacity of 160 mAh g^{-1} at 6 C [57] and 65–130 mAh g^{-1} at 20 C [58, 60], while a much higher capacity of 65 mAh g^{-1} at 50 C for LiFePO$_4$/CNT core-shell hybrid nanowires having 20–30 nm diameter has been reported [59]; however, this is much lower than the LiClO$_4$/CNT cathode prepared by a solution-based two-step method at 80°C (78 mAh g^{-1} at current rate 120 C) [58]. However, all these LiFePO$_4$/CNT core-shell hybrid nanowires showed very stable capacity retention and columbic efficiency. The ultra-high-rate capability and excellent cycling stability achieved can be attributed to the fact that all the LiFePO$_4$ particles are connected electronically by 1D CNTs to form an optimal 3D conducting channel, and CNTs also act as a conductive nanowire. The core-shell hybrid nanostructure ensures the easy transport of Li$^+$-ions as a consequence of its small dimensions and large surface area.

Electrospinning is a simple and versatile method to prepare binder-free electrodes with fibrous morphology and controlled properties. Figure 4.1 shows the schematic illustration of a typical electrospinning setup for preparing metal oxide nanofibers. Toprakci et al. [48] reported that $LiFePO_4/CNT/C$ composite nanofibers were synthesized by a combination of electrospinning and the sol-gel method using CNT as a functional filler and PAN as electrospinning medium and carbon source. The hybrid fibrous electrode shows an initial reversible capacity of 150, 162, and 169 mAh g^{-1} for pristine $LiFePO_4$ powder and $LiFePO_4/C$

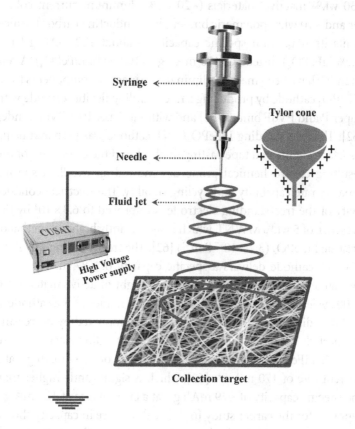

FIGURE 4.1 Schematic illustration of typical laboratory-type electrospinning set up for the production of nanofibrous membranes.

and LiFePO$_4$/CNT/C composite nanofibers, respectively. The LiFePO$_4$/CNT/C cathode exhibited an average reversible capacity of 134 and 121 mAh g^{-1} at 1 and 2 C-rates, respectively, as well as stable cycling stability. The unique fibrous structure, high surface-to-volume ratio, complex porous structure, and shortened Li$^+$-ion diffusion pathway, which enhances the electrode reaction kinetics and reduce the polarization, results in good cycling performance, high reversible capacity, and excellent rate capability [48].

Typically, the electrodes in LIBs contain a considerable amount of inactive material. For example, a LiFePO$_4$ cathode contains ~30 wt.% inactive materials (~20 wt.% aluminum current collector and ~10 wt.% polymeric binder plus conducting carbon), which limits its maximum specific capacity to about 120 mAh g^{-1} (70 wt.% LiFePO$_4$) instead of 170 mAh g^{-1} (100 wt.% LiFePO$_4$). A few attempts have been made to eliminate the inactive components from LiFePO$_4$ cathode by producing a free-standing flexible cathode with Super P and PVdF binder [61] and without Super P or PVdF binder [62]. The free-standing LiFePO$_4$/CNT cathode was prepared using a surface engineered tape casting method, and the evaluation of the resulting electrochemical properties showed an excellent specific capacity rate capability and cycling stability. The electrical conductivity of the free-standing electrode is improved to 6.3×10^2 by the presence of 5 wt.% MWNT, which is significantly higher than that of pristine LiFePO$_4$ (3.7×10^{-7} S m^{-1}) [62]. The free-standing LiFePO$_4$/MWNT cathode delivers a specific capacity of ~87 mAh g^{-1} at a discharge rate of 1905 mA g^{-1} on the weight of active material for 1000 cycles [61]. The electrochemical performance of the cathode at a higher discharge rate (>1700 mA g^{-1}) is improved by decreasing the particle size of LiFePO$_4$ [62]. The free-standing electrode with 90 wt.% LiFePO$_4$ delivered a specific capacity of ~134 mAh g^{-1} at a current rate of 170 mA g^{-1} [62], which is significantly higher than the specific capacity of ~79 mAh g^{-1} at a current rate of 127 mA g^{-1} reported for the earlier study [61]. The difference in capacity shown for the free-standing cathodes prepared using similar methods may be due to the different grade of MWNT used in the different studies.

4.3 SUMMARY

CNTs have shown great potential as composites with metal oxides for improving the performance of LIBs. In recent years, many efforts have been made to fabricate cathodes from CNT. Of the different approaches, the most promising attempts for improving LIB cathode may involve the combination of CNTs with other Li storage compounds with elaborate nanostructure designs. Since $LiCoO_2$, $LiMnO_4$, $LiMn_2O_4$, and lithium metal phosphates are superior among cathode materials for LIBs in terms of capacity, it seems likely than any sufficiently high-capacity cathode hence this chapter more focused on the composite cathode of the aforementioned metal oxides with CNTs. In the composite cathodes, CNTs support the role of conducting additives by forming 3D conducting network structures between the active materials, thereby connecting them together and acting as a mechanical buffer to prevent or minimize the pulverization of electrodes due to the large volume changes during continuous charge/discharge cycling. At present, there is a clear advantage to using CNTs in LIBs as an additive within composite electrodes to increase reversible capacity, enhance rate capability, improve cyclability, and boost electrochemical performance under abused conditions. However, innovations to overcome the current challenges of fast-decaying specific capacity, especially in the initial cycles voltage profiles with CNT composite electrodes will result in specific energy density improvements of more than 50% in full battery capacities. Although the study of CNTs has made much progress, more efforts, research, and energy should be invested for the complete utilization of the potential advantages of CNTs.

REFERENCES

1. Pistoia et al. *J. Electrochem. Soc.* 142 (1995) 2551.
2. Resimers et al. *J. Electrochem. Soc.* 140 (1993) 2752.
3. Li et al. *Solid State Ionics* 67 (1993) 123.
4. Dahn et al. *J. Electrochem. Soc.* 138 (1991) 2207.
5. Koetschau et al. *J. Electrochem. Soc.* 142 (1995) 2906.

6. Jeong et al. *J. Power Sources* 102 (2001) 55.

7. Jin et al. *J. Power Sources* 117 (2003) 148.

8. Kim et al. *J. Power Sources* 119–121 (2003) 686.

9. Suresh et al. *J. Power Sources* 161 (2006) 1307.

10. Suresh et al. *J. Electrochem. Soc.* 152 (2005) A2273.

11. Rodrigues et al. *J. Power Sources* 102 (2001) 322.

12. Suresh et al. *J. Power Sources* 112 (2002) 665.

13. Jin et al. *Electrochem. Commun.* 10 (2008) 1537.

14. Liu et al. *J. Power Sources* 184 (2008) 522.

15. Wang et al. *ACS Nano* 4 (2010) 2233.

16. Xiang et al. *Electrochem. Commun.* 12 (2010) 1103.

17. Sheem et al. *J. Power Sources* 158 (2006) 1425.

18. Zhang et al. *New Carbon Mater.* 22 (2007) 361.

19. Wang et al. *Solid State Ionics* 179 (2008) 263.

20. Sheem et al. *Electrochim. Acta* 55 (2010) 5808.

21. Wang et al. *J. Solid State Electrochem.* 15 (2011) 759.

22. Park et al. *J. Solid State Electrochem.* 14 (2010) 593.

23. Ban et al. *Adv. Mater.* 22 (2010) 145.

24. Lima et al. *Science* 331 (2011) 51.

25. Zhang et al. *Adv. Mater.* 21 (2009) 2299.

26. Luo et al. *Adv. Mater.* 24 (2012) 2294.

27. Yan et al. *J. Mater. Chem. A* 5 (2017) 4047.

28. Li et al. *Electrochem. Solid State Lett.* 9 (2006) A126.

29. Li et al. *Carbon* 44 (2006) 1334.

30. Huang et al. *Scr. Mater.* 64 (2011) 122.

31. Huang et al. *J. Mater. Chem.* 21 (2011) 10777.

32. Ban et al. *Adv. Energy Mater.* 1 (2011) 58.

33. Mun et al. *J. Mater. Chem. A* 2 (2014) 19670.

34. Fang et al. *Nano Energy* 12 (2015) 43.

35. Liu et al. *J. Power Sources* 195 (2010) 4290.

36. Tang et al. *Electrochim. Acta* 166 (2015) 244.

37. Jia et al. *Chem. Commun.* 47 (2011) 9669.

38. Lou et al. *J. Mater. Chem. A* 1 (2013) 3757.

39. Reddy et al. *Nano Lett.* 9 (2009) 1102.

40. Padhi et al. *J. Electrochem. Soc.* 144 (1997) 1188.

41. Bramnik et al. *J. Solid State Electrochem.* 8 (2004) 558.

42. Wang et al. *J. Electrochem. Soc.* 154 (2007) A1015.

43. Mohamed et al. *Int. J. Electrochem.* 283491 (2011) 1.

44. Kim et al. *J. Adv. Eng. Technol.* 1 (2008) 107.

45. Zhou et al. *Chem. Commun.* 46 (2010) 7151.

46. Yang et al. *J. Mater. Chem.* 22 (2012) 7537.

47. Gananavel et al. *J. Electrochem. Soc.* 159 (2012) A336.
48. Toprakci et al. *ACS Appl. Mater. Interfaces* 4 (2012) 1273.
49. Murugan et al. *Inorg. Chem.* 48 (2009) 946.
50. Schneider et al. *Eur. J. Inorg. Chem.* 28 (2011) 4349.
51. Qin et al. *J. Power Sources* 248 (2014) 588.
52. Qina et al. *Electrochim. Acta* 115 (2014) 407.
53. Adepojua et al. *ECS Trans.* 80 (2017) 267.
54. Wang et al. *Adv. Energy Mater.* 6 (2016) 1600426.
55. Wu et al. *Adv. Energy Mater.* 3 (2013) 1155.
56. Qiao et al. *Electrochim. Acta* 232 (2017) 323.
57. Kim et al. *Chem. Commun.* 46 (2010) 7409.
58. Jegal et al. *J. Power Sources* 243 (2013) 859.
59. Yang et al. *J. Mater. Chem. A* 1 (2013) 7306.
60. Wang et al. *J. Power Sources* 291 (2015) 209.
61. Susantyoko et al. *J. Mater. Chem. A* 5 (2017) 19255.
62. Susantyoko et al. *RSC Adv.* 8 (2018) 16566.

CHAPTER 5

Carbon Nanotube/ Polymer Nanocomposite Electrolytes for Lithium Ion Batteries

5.1 INTRODUCTION

Batteries are electrochemical storage systems that are basically comprised of three components: an anode, a cathode, and an electrolyte. The performance of lithium ion batteries (LIBs) greatly depends on the electrochemical reactions taking place in the electrolyte [1]; hence, electrolytes are the critical component for high-performance LIBs with a long lifetime. An electrolyte acts as a medium for the transportation of Li$^+$-ions and also as an interface between two electrodes [2]. During the charge/discharge cycle, Li$^+$-ion transportation involves the migration and diffusion of ions between the electrodes through the electrically insulating electrolyte; that is, the electrolyte is an ionically conducting but an electronically insulating material [3]. The electrolyte must have good electrochemical stability and compatibility with electrode materials [4, 5]. The primary requirements for an electrolyte in LIBs are as follows:

1. It must be a chemical medium having Li^+-ion species.

2. It should have good ionic conductivity.

3. It should be electrically insulating.

4. It should be ionically conducting.

5. It should have good electrochemical stability.

6. It should be thermally stable.

7. It must be economically viable.

8. It should be environmentally friendly.

Based on their physical properties, electrolytes for LIBs are classified as solid, liquid, and gel electrolytes, each having its own advantages and limitations. Among the different electrolytes, liquid electrolytes are most commonly used in LIBs. A liquid electrolyte is a homogeneous blend of organic carbonates and Li salt [6]. Widely used organic carbonates are ethylene carbonate (EC), dimethyl carbonate (DMC), diethyl carbonate (DEC), and ethyl methyl carbonate (EMC) or its mixture along with the Li salt [1, 7]. Owing to their efficient ion transport capability and transference number across the two electrodes, liquid electrolytes typically show an ionic conductivity of 5–10 mS cm^{-1} at ambient temperature and usually operate in a temperature range of –30°C–60°C. They have, however, many limitations, including issues of storage and transportation, leakage, solid electrolyte interface (SEI) layer formation at electrode–electrolyte interfaces, difficulty in handling and processing, and environmental issues. Even though these liquid electrolytes improve LIB performance, their high vapor pressure and low flash point represent a major safety concern [2, 8, 9]. Hence, LIBs comprising liquid or conventional gel polymer electrolytes (GPEs) carry a serious safety risk. In 2006, the explosion of Sony laptop triggered a recall of almost six million Li^+-ion packs. A series of battery explosions have been reported recently, including by major electronics firms such as Dell, HP, and Samsung. Fires originating in the Samsung Galaxy Note 7 turned a dream phone into a nightmare. The main-ship battery in the Boeing 787 Dreamliner and MH 370 again demonstrates the safety issues involved in the use of LIBs. Samsung, the giant in portable electronics devices, officially revealed the reason for the explosion of the Galaxy Note 7 as a short-circuit problem caused by the contact of two electrodes due to the inefficiency of the separator. The flammability of conventional electrolytes paved the way for the development of better alternatives with good thermal and mechanical stability.

Solid electrolytes have emerged as safer alternatives that offer significant benefits, including higher thermal and electrochemical stability, suppression of dendrite growth, and thin film manufacturability, leading to tantalizing applications like printable, flexible, and stretchable batteries. However, the low ionic conductivity of solid electrolytes, especially at room temperature, remains a challenge and limits their practical applications at room temperature. Depending on the matrix system, solid electrolytes are classified as ceramic electrolytes and polymer electrolytes (PEs). Compared to ceramic solid electrolytes, PEs have higher ionic conductivity, are easy to process, and are flexible.

To tackle the disadvantageous properties of solid and liquid electrolytes, a new class of electrolytes called gel polymer electrolytes (GPEs) have been developed which combine the advantageous properties of liquid and solid polymer electrolytes (SPEs). GPEs are very similar to conventional liquid electrolytes. These electrolytes reduce the use of carbonate solvents, leading to improved processability and safety by eliminating the problems associated with the flammable nature of solvents.

The excellent processability and flexibility of these electrolytes help in the fabrication of ultrathin Li^+-ion cells with various geometries [10, 11].

The high crystallinity of the host polymer matrix is one of the major factors contributing to the low ionic conductivity of PEs. The ionic conductivity, transport number, and thermal stability of SPEs can be marginally improved by the incorporation of ceramic nanofillers (e.g. Al_2O_3, SiO_2, $BaTiO_3$, γ-$LiAlO_2$, ZrO_2, BN, TiO_2, aluminates, and titanates) [2, 12] and carbon nanomaterials (e.g. carbon nanotubes [CNTs] and graphene) without compromising their mechanical properties, while solvents like ionic liquids or blends of ionic liquids with carbonate solvents may be used to improve the safety of LIBs [13]. Even though the amount of these nanostructures incorporated into the polymer matrix are very low, they can profoundly affect the mechanical strength and ionic conductivity of PEs. The enhancement in ionic conductivity is attributed to a combination of factors, including the role of nanofillers as solid plasticizers [14] and electrolyte dissociation promoters [15]. The nanoparticles interact with polymer chains and electrolyte through Lewis acid–base interactions, which reduce the crystallinity of the host polymer and improve the electrolyte dissociation. In addition, the nanostructures also enhance interfacial stability at the PE–Li electrode interface and improve the cycling stability of the cell [16].

CNT is an allotrope of carbon with unique structural features and has a large aspect ratio, high strength, and modulus, making it an excellent reinforcing agent for a polymer matrix [17]. Compared to other allotropes of carbon, CNTs and graphene are well explored as electrode materials for the development of high-performance LIBs; however, there has been little research into the electrochemical properties of CNTs as electrolytes due to the fact that they are electronically conducting and thus can pose the risk of electrical shorting. Only a few studies have been reported on the use of CNTs as a functional filler for the development of electrolytes for LIBs. By incorporating CNTs into the electrolyte, the filler/matrix interface gives a low energy path for Li^+-ion transportation [18]. The salt dissociation of Li salts and ion transport is also facilitated through CNT's rich electron cloud [19].

5.2 POLYETHYLENE OXIDE/CNT COMPOSITE ELECTROLYTE

In PEs, the polymer materials act as a host matrix that is activated by conventional liquid electrolyte. Some of the generally used polymer host matrices include poly(vinylidene fluoride) (PVdF); its copolymer with hexafluoropropylene, P(VdF-co-HFP); poly(acrylonitrile) (PAN); poly(ethylene oxide) (PEO); and poly(methyl methacrylate) (PMMA). [20]. Their mechanical properties are the main concern; that is, the lesser amount of polymer in GPEs limits their practical application [20–22]. Amongst them, PEO and PVdF have been widely studied as matrices for CNT-based SPEs and GPEs, respectively [23, 24]. The ether linkages in PEO lead to local relaxation and segmental motion of the polymeric chains, which facilitates Li^+-ion conduction because of its ability to make coordination with alkali metal ions when it is used as a PE in LIBs [11, 23]. PEO is a highly crystalline polymer with a melting temperature (T_m) of 67°C and glass transition temperature (T_g) of –65°C. It is widely accepted that PEO can form complexes with a large number of Li salts and transport Li^+-ion in SPEs. Thus, significant ion conduction occurs in the amorphous phase where the conductivity is two or three orders of magnitude higher than in the crystalline region. However, PEO often shows high crystalline ratios in sub-ambient temperature regions, and its ionic conductivity is lower than 10^{-5} S cm^{-1} due to its particular molecular structure [14, 25]. The decrease in crystallinity and T_g can result in an increase in conductivity, which can be achieved by modifying the solvating polymer, by forming composites with ceramic or carbon nanostructures, by blending with other polymers, or by plasticizing with low-molecular-weight materials. Therefore, many researchers have dedicated their efforts to develop PEO-based SPEs with high ionic conductivity at low temperatures, especially below room temperature, and a high Li$^+$-ion transport number toward Li metal anodes, as well as good mechanical and electrochemical properties [14]. Among the different polymers popularly used for the preparation of SPE, PEO is the most studied polymeric material for the CNT-based composite electrolyte system [19, 26].

Ibrahim et al. [26] reported a new composite GPE based on PEO incorporated with amorphous CNT (α-CNT), prepared via the chemical route at low temperatures. The composite GPE was prepared using a solution-casting technique, using PEO as host matrix, LiPF$_6$ as the Li salt, EC as the plasticizer, and α-CNTs as the filler. The α-CNTs loading was kept constant at 5 wt.%. The study shows that the incorporation of 15 wt.% EC and α-CNTs into the salted polymer showed a significant conductivity increase from 10^{-5} to 10^{-3} S cm^{-1} compared to the electrolyte with 20 wt.% LiPF$_6$. The pronounced enhancement in the ionic conductivity of the electrolyte by the addition of highly flexible α-CNTs into PEO networks is primarily attributed to the improved interaction and crosslinking between α-CNTs and PEO molecules. The composite electrolyte is reported to have an ionic conductivity of 1.3×10^{-3} S cm^{-1}, which is about 530% higher than the electrolyte without α-CNTs [26]. The study shows that the conductivity of PEO (3×10^{-10} S cm^{-1}) increases by five orders of magnitude upon the addition of LiPF$_6$ (4×10^{-5} S cm^{-1}) and four orders of magnitude upon the addition of EC (2×10^{-4} S cm^{-1}). The sudden increase in conductivities is attributed to the presence of Li$^+$-ions in the PEO and the increase in the flexibility of the polymer chains due to the plasticizing effect of EC and the lowering of the crystallinity of the PEO by α-CNTs on the GPEs.

When EC was added into the system, more salts were dissociated into ions, which have a low viscosity and therefore increasing ionic mobility. The addition of α-CNTs increases conductivity by inhibiting the recrystallization of the PEO chains and providing a Li$^+$-ion conducting pathway at the filler surface [27]. Under the influence of a potential gradient, the coordination spheres of the

Li$^+$-ions keep changing to adjacent locations, and this is assisted by the segmental motion of the polymer chain. The polymer chain undergoes reorganization during the continuous breaking and reforming of the coordination sphere for Li$^+$-ions [28].

5.3 POLYVINYLIDENE DIFLUORIDE/ CNT COMPOSITE ELECTROLYTE

PVdF and its copolymer P(VdF-*co*-HFP) have been widely studied as host polymers for preparing PEs. These fluoropolymers have received great attention due to their good electrochemical stability, high dielectric constant (~8.4), affinity to electrolyte solution, and presence of strong electron-withdrawing fluorine, making them a good candidate for the preparation of an electrolyte-cum-separator for LIBs. GPE based on these polymers shows high ionic conductivity (in the range of 10^{-4}–10^{-3} S cm^{-1} at room temperature) and good electrochemical stability [24, 29]. Also, PVdF possesses many remarkable properties, such as good thermal stability under operating and processing temperature, non-flammability, and excellent chemical resistance in combination with very low creep and high mechanical strength. Unfortunately, pristine PVdF is partially soluble in the organic liquid electrolyte that is generally used for the preparation of GPEs. This could result in the loss of mechanical strength and may cause internal short-circuiting, leading to cell failure. The crystalline part of PVdF hinders the migration of Li$^+$-ions, and hence batteries with GPEs based on PVdF show lower charge/discharge properties and poor rate capability. Recently, many groups have reported that these problems can be resolved by the addition of nanosized functional materials as fillers [30, 31].

Kan et al. [32] made an attempt to study the effect of the CNTs on ionic conductivity and their compatibility with the composite GPEs based on P(VdF-*co*-HFP). The nanocomposite membrane was prepared by a solution casting process using P(VdF-*co*-HFP) with varying concentration of CNT (0%–5%). With the incorporation of CNT into the P(VdF-*co*-HFP), 3D porous structures are formed, which increases the electrolyte uptake of the membrane from 140% for pristine P(VdF-*co*-HFP) membrane to 220% with 3% loading of CNT, and on further increase in CNT loading, the electrolyte uptake of the membranes decreases, indicating the agglomeration of nanostructures.

The electrolyte uptake of the composite membrane greatly depends on its microporous structure and affinity to the electrolyte. The presence of hydrophobic CNTs helps to form uniform 2D porous structures, which result in the electrolyte absorption capability of the composite membrane. The uniform microporous structures of composite membranes provide more pore volume and a large surface area, which in turn results in stronger affinity to the electrolyte. A filler loading of 3% CNT into the P(VDF-*co*-HFP) shows the highest ionic conductivity, that is, 4.4×10^{-3} S cm^{-1}, which is about 75% higher than that of the microporous membrane prepared with pristine P(VdF-*co*-HFP). The electrolyte prepared without CNT and loaded with 3% CNT showed an initial interfacial resistance of 392 and 278 Ω respectively against Li/Li metal electrodes. The higher ionic conductivity for CNT-loaded samples is due to the interconnected pore structure of the membranes. At a higher loading of CNT, that is, >3%, the interconnection of the pores may be broken due to the agglomeration of CNTs. Linear sweep voltammograms of the cells also showed a similar trend, and the electrolyte with 3% CNT showed an anodic stability of 5.4 V while pristine P(VdF-*co*-HFP) showed an anodic stability of only 5 V. The higher ionic conductivity and increased anodic stability is ascribed to the excellent affinity of the membrane to the electrolyte, which partially dissolves the amorphous regions of the polymer chains.

Charge/discharge cycling performance of 3% CNT-loaded P(VdF-*co*-HFP) electrolytes were evaluated by galvanostatic charge/discharge cycling in the cell configuration of Li/GPE/ LiFePO$_4$ at a rate of 0.1 C between 2.7 and 4.3 V. The cell shows an initial charge/discharge capacitance of 140 mAh g^{-1}, which is 82.4% of the theoretical capacity of LiFePO$_4$. After 30 cycles, the cell retains a capacitance about 97% of its initial capacitance, that is, 136 mAh g^{-1}, which shows the suitability of GPE-comprising CNTs in LIBs. Compared to pristine P(VdF-co-HFP)-based electrolytes, GPEs incorporated with CNTs have better electro-chemical stability, suggesting the synergistic effect of CNTs and P(VdF-co-HFP). The presence of CNTs in the electrolyte improves its ionic conductivity, anodic stability, and interfacial properties as well as lowering the crystallinity of the composite electrolyte due to the inherent electrochemical and thermal properties of CNTs.

5.4 POLYVINYLIDENE DIFLUORIDE/CNT/ CLAY HYBRID ELECTROLYTE

CNT and nanoclay have been used as high-performance fillers for polymers in various applications; however, the use of CNTs is limited in PEs due to their electronic conductivity, which poses a great risk of short-circuiting. There have been many reported studies on the use of polymer clay nanocomposite or polymer CNT nanocomposites as high-performance electrolytes for LIBs, and even of clay mixed with Li salt and plasticized with room temperature ionic liquids as an electrolyte for an energy stor-age device that can operate at temperatures as high as 200°C [2]. Ajayan et al. [19] demonstrated a novel method for using CNTs as a high-performance ion conducting promoter-cum-reinforcing filler in PEs. In the study, they showed that nanotubes can be pack-aged within insulating clay layers to form effective 3D nanofillers. The PE incorporated with 10% clay–CNT nanofiller was reported to have an ionic conductivity of 2×10^{-5} S cm^{-1}. The demonstra-tion of CNTs hybridized within insulating clay platelets opens up the avenues to the utilization of CNTs and graphene in PEs for enhancement of various critical properties.

The 1D nanotubes are grown within 2D montmorillonite clay platelets to create hybrid 3D nanofillers [33–35]. The clay is first treated with Fe(NO$_3$)$_3$ to form clay-supported iron oxide, and CNTs are grown within the ceramic layers using a chemical vapor deposition (CVD) method. The SPE was prepared by a solution casting method using PEO as the matrix, LiClO$_4$ as Li salt, and clay–CNT 3D nanostructure as the ionic conductivity promoter. The filler content was varied from 0% to 15%; however, the electrolyte with 10% clay–CNT exhibited the highest ionic conductivity value. The increasing trend of ionic conductivity with filler loading is due to the combination of salt dissociation and ion mobility. Upon further increase in the clay–CNT content above 10%, the ionic conductivity shows a slight decrease, indicating the potential aggregation of hybrid fillers impeding the Li$^+$-ion transport in PEO [36]. In the electrolyte system, positive Li$^+$-ions exhibit strong affinity to the high negative charge of the electron cloud in the outer surface of CNTs as well as the negative oxygen atoms on the clay, which can result in the separation of the contact ion pairs [37]. Also, the PEO chains can be unfolded, leading to an increase in free volume due to the high aspect ratio of the 3D hybrid nanoscale fillers. An increase in free volume has been associated with the greater mobility of ions due to least resistance paths in the polymer, which facilitate the easy transportation of Li$^+$-ions, leading to higher ionic conductivity. The impedance spectroscopy of the PEO filled with clay–CNT without Li salt showed a resistance value above 10^8 Ω, indicating good electrical insulation properties. This high insulation properties may be due to the fact that clay platelets attached to CNTs prevent electron conduction in CNTs, which, in turn, prevents electrical shorting. While the CNTs are growing within the clay layers, the clay gets partially exfoliated, resulting in a high-aspect-ratio hybrid filler, which will act as a barrier for electron transportation by clutching it with surface charges on the clay platelets. An increase in tensile strength of up to 160% and improved thermal stability was observed in clay–CNT hybrid filled PEO electrolyte [19].

5.5 SUMMARY

In summary, among different electrolyte systems such as liquid electrolytes, gel electrolytes, and solid electrolytes, SPEs have unique properties such as safety at high temperatures, structural integrity, dimensional stability, easy processability, and good shelf life, along with easy tunability of ionic conductivity and electrochemical properties. In comparison with other polymeric matrices, PEO and PVdF have been well studied as composite SPEs with CNT. The incorporation of a small fraction of CNTs in PEO or PVdF complexed with Li salt significantly improves ionic conductivity and Li^+-ion transportation levels, as well as thermal and mechanical properties which profound in charge/discharge cycling stability and rate capability of the LIBs. The presence of CNTs facilitates the chain flexibility of the polymer, leading to segmental motion due to the lowering of crystallinity and activation energy in the Li^+-ion conduction. The reports on polymer/CNT composite electrolytes open up a new avenue for the utilization of the unique electrochemical properties of carbon nanomaterials for the development of safe electrolytes for high-performance LIBs.

REFERENCES

1. Kang. *Chem. Rev.* 114 (2014) 11503.
2. Prasanth et al. *Electrochim. Acta* 55 (2010) 1347.
3. Amereller et al. *Solid State Chem.* 42 (2014) 39.
4. Tarascon et al. *Nature* 414 (2001) 359.
5. Hammami et al. *Nature* 424 (2003) 635.
6. Yoshio et al. *Lithium-Ion Batteries: Science and Technologies.* New York: Springer Science and Business Media; 2009.
7. Xu. *Chem. Rev.* 104 (2004) 4303.
8. Padmaraj et al. *J. Phys. Chem. B* 119 (2015) 5299.
9. Aydın et al. *J. Mater. Res.* 29 (2014) 625.
10. Ahmad. *Ionics* 15 (2009) 309.
11. Fenton et al. *Polymer* 14 (1973) 589.
12. Lim et al. *J. Nanosci. Nanotechnol.* 18 (2018) 6499.
13. Armand et al. *Nat. Mater.* 8 (2009) 621.
14. Croce et al. *Nature* 394 (1998) 456.
15. Sun et al. *J. Electrochem. Soc.* 147 (2000) 2462.
16. Jiang et al. *J. Power Sources* 141 (2005) 143.
17. Coleman et al. *Carbon* 44 (2006) 1624.
18. Udomvech et al. *Chem. Phys. Lett.* 406 (2005) 161.

19. Tang et al. *Nano Lett.* 12 (2012) 1152.
20. Manuelstephan. *Eur. Polym. J.* 42 (2006) 21.
21. Ahmad. *Ionics* 15 (2009) 309.
22. Fergus. *J. Power Sources* 195 (2010) 4554.
23. Scrosati et al. *MRS Bull.* 25 (2000) 28.
24. Pasquier et al. *Solid State Ionics* 135 (2000) 249.
25. Meyer. *Adv. Mater.* 10 (1998) 439.
26. Ibrahim et al. *Solid State Commun.* 151 (2011) 1828.
27. Saikia et al. *Electrochim. Acta* 54 (2009) 1218.
28. Nookala et al. *J. Power Sources* 111 (2002) 165.
29. Prasanth et al. *J. Power Source* 186 (2011) 6742.
30. Prasanth et al. *J. Power Source* 187 (2008) 437.
31. Prasanth et al. *J. Power Source* 267 (2014) 48.
32. Kan et al. *Int. J. Chem. Eng. Appl.* 4 (2013) 42.
33. Zhang et al. *Adv. Mater.* 18 (2016) 73.
34. Wang et al. *Mater. Sci. Eng.* 490 (2008) 481.
35. Zhao et al. *Polym. Compos.* 30 (2009) 702.
36. Baxter et al. *Energy Environ. Sci.* 2 (2009) 559.
37. Sim et al. *Spectrochim. Acta Part A* 76 (2010) 287.

CHAPTER 6

Graphene and Its Composites as Anodes for Lithium Ion Batteries

6.1 INTRODUCTION

In lithium ion batteries (LIBs), materials that can store enough Li^+-ions during charging are generally used as anodes. In Whittingham's battery, metallic Li was used as the electrode. Because of safety-related issues and poor rechargeability, 6 mAh Li metal anodes are commercially unattractive, even though they demonstrate a very high capacity of 3,860 mAh g^{-1} [1]. As a result, researchers focused on finding alternatives to metallic Li in LIBs, and a "Li^+-ions rocking-chair," "Li^+-ions swing," or "Li^+-ions" concept of cell-operation battery utilizing an intercalation compound as the anode for the Li secondary battery was proposed. In 1980, Rachid Yazami [2] successfully demonstrated the reversible electrochemical intercalation of Li in graphite as an anode in LIBs [2]. The use of the intercalated carbon/graphitic anode solved the problem of poor Li metal rechargeability due to the formation of dendrites and mossy Li metal deposits with only a fairly small voltage penalty (0.050 mV), thereby greatly increasing the safety of high-energy LIBs. A wide range of carbon materials have been investigated as anodes in LIBs and show good performance as the host for Li^+-ions [3–8]. Recently, other high-capacity alternative anode materials such as Al, Sn, Si, Bi, Ti, and Sb have been demonstrated as high-capacity anodes in LIBs [9–13]. However, these metal oxides suffer mechanical instability due to large volume expansion/contraction when alloyed and de-alloyed reactions with Li^+-ions lead to the mechanical cracking of the electrode and thus poor cyclability [14–16]. Many approaches have been proposed to address these problems, including (i) reducing the size of the particles, (ii) applying pressure to cells, (iii) using elastomeric binders, or (iv) forming a composite with conductive materials such as conducting polymers [17–20], carbon nanofibers (CNFs) [21], carbon nanotubes (CNTs) [3, 22, 23] and graphene [5, 22]. The small particle size, high conductivity, and large electrochemically accessible surface area of CNT and graphene composites make them an ideal anode material for high energy LIBs. Among different anodic materials, this chapter focuses only on graphene and its composite with SnO_2 as the best representatives of metallic and non-metallic anode materials. It discusses the different mechanisms and approaches proposed to address the problems associated with their electrochemical properties.

6.2 GRAPHENE AS ANODE MATERIAL

Increasing the theoretical capacity of commercially used graphite for the anode has been the key issue for developing efficient LIBs [24–27]. Many innovative methods have been proposed to increase the capacity of anodes in LIBs over the years. For instance, using a simple method involving the sonication and ultracentrifugation of graphite [28–30], graphene ink [31–33] and Cu-supported graphene nanoflakes [34] were used as an anode in LIBs. The studies showed a high specific capacity of 164 mA g^{-1} for an energy density of 190 Wh kg^{-1}. The high capacity was obtained by cleverly balancing the cell composition and controlling the initial irreversible capacity in the anode, which led to stable operation of the battery for over 80 charge/discharge cycles. The key was to control the morphological properties of the graphene nanoflakes to achieve large Li uptake by the designed anode. Li$^+$-ions could be absorbed not only on both sides of graphene sheets, but also on their edges, defects, vacancies, and covalent sites, which would be very favorable for application in LIBs [35, 36]. In their studies using graphene nanosheet (GNS) for the anode, Vargas et al. [37] reported a high capacity of 2000 mAh g^{-1} with a subsequent drop that stabilized at 600 mA g^{-1} with a working voltage of 0.9 V. Electrolyte decomposition with solid electrolyte interface (SEI) was attributed to this drop. The stabilized value at 600 mAh g^{-1} was still high compared to the theoretical value due to the high surface area offered by GNS [37–39].

Varying the degree of oxidation in the preparation of graphene sheets from graphene oxide (GO) [40, 41] has been shown to have a huge effect on the final quality of the GNS. A higher degree of oxidation resulted in thinner GNS with a high specific area, which is desirable for the diffusion of Li$^+$-ions [42]. When used as an anode material, ultrathin graphene sheets showed a high reversible capacity due to high Li absorption [43–47]. Doping has been recognized as an easy way to improve the performance of GNS as an anode material [48, 49]. In 2010, Bhardwaj et al. [50] compared the electrochemical properties of graphene nanoribbon (GNR) and oxidative GNR (ox-GNR) and reported that ox-GNR performs best. The electrode fabricated by ox-GNR showed much better electrochemical performance in LIBs with reversible capacities in the range of 800 mAh g^{-1}, with very stable cycling performance and rate capability. The presence of a high amount of hydrogen and oxygen atoms on the graphene surface probably triggered more Li$^+$-ion intake [50]. This development immediately points to the possibility of doped or functionalized graphene nanostructures as anodes in LIBs. Leela et al. [51] synthesized an N-doped graphene directly on a Cu foil through a liquid precursor-based chemical vapor deposition (CVD) method, which offers almost double the reversible capacity as compared to a pristine graphene anode [51]. The enhanced specific capacity of the N-doped graphene anode was ascribed to the excess number of defect sites created during N-doping, which enhances the Li$^+$-ion insertion capacity of GNS. A similar observation has been reported by Wang et al. [52] and Cai et al. [49] on N-doped GNS. N-doped graphene anodes, prepared by high-temperature treatment of GO in an ammonia atmosphere [52] or N-doped and B-doped graphene anodes prepared through oxidation and thermal exfoliation of natural flaky graphite powder [48] were also reported to have an excellent specific capacity and rate capability. The N-doped and boron-doped GNS (B-doped GNS) anode have shown specific capacities of 199 and 235 mAh g^{-1}, respectively, even at a very high current rate of 25 A g^{-1}. The drastically enhanced electrochemical behavior of doped graphene structures has been attributed to the synergetic effect of their unique 2D network structure, disordered surface, heteroatomic defects, improved wettability between electrode and electrolyte, increased interlayer spacing, higher electrical conductivity, and thermal stability. The disordered structure, heteroatomic defects, and level of doping introduce more defect sites for Li insertion during charge/discharge cycling, while wettability, high thermal stability, and electronic conductivity contribute to the improvement of the structural stability and kinetics of Li$^+$-ion and electron transfer. The improved stability and fast ion transfer kinetics result in the rapid absorption of a large amount of Li$^+$-ions on the surface of the anode and the ultrafast diffusion of ions to the interior of the electrode. The improved electronic conductivity ensures rapid electron transport from the active material in the anode to the current collector, leading to excellent specific capacity and rate capability.

Cai and co-workers [49] synthesized N-doped graphene by high nitrogen loading of 7.04% by annealing pristine graphene sheets with melamine, and the N-doped electrode showed a very high initial reversible capacity of 1123 mAh g^{-1} at a current of 50 mA g^{-1}. A high stable capacity of 241 mAh g^{-1} could still be observed at a high current density of 20 A g^{-1}. The N-doped graphene anodes are the predominance of the pyridinic type of nitrogen atoms over pyrrolic and graphitic types. The existence of a 2D structure, pyridinic nitrogen atoms, and the doping level were all thought to be responsible for the excellent electrochemical performance of the battery [53]. Thus, ensuring the presence of a maximum amount of pyridinic nitrogen atom is a major factor in attaining high specific capacity for N-doped graphene electrodes. The doping of graphene with nitrogen has always been restricted to approximately 10 wt.% due to the tendency toward structural instability caused by nitrogen atoms on the 2D honeycomb lattice of graphene [54]. However, utilizing a metal organic framework, Zheng et al. [55] showed that it is possible to dope graphene with more than 10% of nitrogen. In their research, they reported a high value of 17.72 wt.% N-doped graphene by pyrolysis of a zeolitic imidazole framework containing nitrogen at 800°C under nitrogen atmosphere. When used as an anode for LIBs, a high capacity of 2132 mAh g^{-1} at 50 cycles with a current density of 100 mA g^{-1} was maintained, while a capacity of only 786 mAh g^{-1} remained after 1000 cycles at a current rate of 5 A g^{-1} [55–58]. A comparison of doping materials was explored by using nitrogen and boron [48] as dopants in graphene sheets, which was later used as anode material for the LIBs. It was shown that N- or B-doped graphene anodes exhibited high power and energy density when subjected to high-rate charge and discharge conditions. Further, both B- and N-doped graphene sheet electrodes showed a desirable stability and high capacity at very high current rates with no capacity fluctuations, a quality generally observed with un-doped graphene sheets. The disordered morphology, increased intersheet distance, heteroatomic defects, and better electrode wettability induced on the graphene structure by nitrogen and boron allowed rapid and ultrafast surface absorption of the Li^+-ions, leading to an improvement in thermal and electrical conductivity. Reduced graphene sheets with a certain amount of residual oxygen were also reported as a superior anode in LIBs [59]. In addition to doping, many methods involving the modification of graphene as an anode material have been attempted by researchers. One notable modification is the use of graphene paper electrodes [60, 61]. In a simple method of freeze-drying, aerogel of GO was obtained, which was later treated in air and mechanically pressed to achieve graphene folded sheets [61]. The folded graphene sheets showed a coulombic efficiency of 79.2% in the first charge/discharge cycle with an increase to 98% in the second cycle. These values can only be achieved if the reversible capacity of the graphene paper [62–64] is very high. The folded structure of the sheets in several layers provided increased Li ion insertion active sites, for example, nanopores and edge-type sites, leading to high Li absorption. In a further modification of the graphene paper sheets, flexible holey graphene paper electrodes showed high performance as an electrode material in LIBs [60]. Inaccessible volume resulting from the compact geometry of the graphene stacks led to difficulty in ion mobility, resulting in a mediocre performance by the graphene electrode. Increasing the in-plane porosity [65, 66] is a critical step to increasing the volume accessible by the Li^+-ions. In-plane porosity was achieved in the basal planes by using a wet chemical method in combination with ultrasonication in the presence of mild acid oxidation [60]. The introduced porosity led to abundant ion binding sites, excellent high-rate Li^+-ion storage, enhanced ion diffusion and high performance in the battery [60].

Three main strategies are widely used to improve the energy performance of graphene, namely the creation of defects [67, 68] and the doping and restacking of GNS [69]. In achieving high performance in terms of Li^+-ion storage capabilities, it has been shown that the synergistic effects of the above strategies can lead to long-term and ultrafast cycling capability. Wang et al. [70] utilized an in-situ method to fabricate a hierarchical porous structure

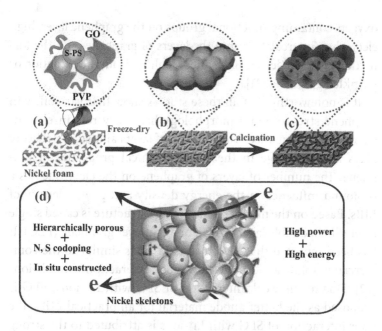

FIGURE 6.1 Schematic illustration of the synthesis procedures of the doped hierarchically porous graphene (DHPG) electrodes: (a) impregnating the precursor solution containing GO, poly(vinyl pyrrolidone) (PVP), and S-PS into the porous collector of nickel foam; (b) freeze-drying the precursor solution; (c) in-situ calcining the precursor gel; (d) illustration of the features in the interior structure of DHPG electrodes. Reproduced from Wang et al. [70].

with a high conducting network. They used a doped heteroatom graphene as one electrode. Figure 6.1 shows the schematic of the method used to fabricate the porous graphene electrode. The hierarchical porous structured anode exhibited high surface area and porosity. The electrode was able to deliver power density as high as 116 kW kg^{-1} with a very high energy density of 322 Wh kg^{-1} at 80 A g^{-1}, comparable to the power density of a supercapacitor. In addition, the electrode showed a long cycling capability with no appreciable loss over 3000 cycles across a wide temperature range of −20°C–55°C. The high irreversible capacity achieved in graphene anodes is mainly due to the reaction of Li$^+$-ions with oxygen-containing functional groups on the graphene layer, high electrolyte intercalation into the layers of graphene, and the formation of SEI in defects or cavities resulting from the oxidation or wrinkled graphene [71].

It is noteworthy that all these studies show that variability in graphene structure and morphology has a key role in electrochemical performance. The number of layers and edge structure plays a critical role in the electrochemical properties of graphene. The number of layers of graphene on the electrode has a profound influence in the energy density and power density of LIBs. Based on the number of layers, the structure is called single layer (SLG), double layer (DLG), or a few layer graphene (FLG). It has been reported that FLG inserts Li$^+$-ions similar to the commercial graphite anode and shows good interaction with Li$^+$-ions [72]. Due to the weak interaction of SLG with Li$^+$-ions, FLG is proposed as the better anode material for all practical LIBs. The poor interaction of SLG with Li$^+$-ions is attributed to the strong coulombic repulsion between Li$^+$-ions on opposite surfaces of the GNS. However, Valota et al. [73] reported that SLG has higher charge rate constants than FLG and better ion transfer kinetics, leading to higher energy and power density [73]. Uthaisar et al. [74], using simulation, studied the effect of the edge structure of graphene on Li$^+$-ion diffusion through different types of edges, and they predicted that edges of the graphene have a much lower energy barrier to Li$^+$-ion diffusion compared to their bulk counterparts. Hence, graphene nanostructures with a greater number of edges would be expected to show higher specific capacity. However, the nanostructures have a number of active sites that favor the formation of a thick SEI layer due to a side reaction with electrolyte, leading to a fast drop-off in specific capacity in charge/discharge cycling. It has been reported that the diffusion of Li$^+$-ions has a preferential path inside the graphene sheets [75], and it has been theoretically shown that Li$^+$-ions preferentially diffuse through a favorable path along the C–C bond axis of the GNS compared to the midpoint of C–C bonds.

6.3 COMPOSITE ANODE OF GRAPHENE

Recently, there have been attempts to combine graphene with other materials to make composite materials for an anode with improved performance [66, 76]. In their review on the nanostructured materials for LIBs, Ji et al. [71] observed enhancement in the electrochemical performance of LIBs fabricated with graphene and metal/metal oxide composite anode due to various factors, high Li^+-ion storage of both graphene and nanoparticles of metal/ metal oxide, and good electrical conductivity of the graphene, ensuring excellent conductivity to adjacent particles. In addition, restacking of GNS is prevented by the confinement of metal oxide or metal nanoparticles. Various different metal/metal oxide composites with graphene have been studied as anode materials in LIBs. The results of tests on composites of SnO_2 [77], TiO_2 [78], Fe_2O_3 [79], Fe_3O_4 [80], CoO [81], V_2O_5 [82], VO_4 [83], MoS_2 [84] and so on with graphene have been reported. As the basic mechanism and approaches are similar in all these materials, this section focuses only on SnO_2/graphene composites. The section also covers the effect of modification, morphology, and binary and ternary composites on the electrochemical properties of SnO_2 anodes for LIBs.

6.3.1 SnO₂/Graphene Composite Anode

Among most anode materials for LIBs, metallic Sn possesses a relatively high theoretical capacity (994 mAh g^{-1}) [85] and is considered to be a promising alternative to graphite anodes (372 mAh g^{-1}) for the development of next-generation LIBs with high energy and power density [86]. Apart from their high theoretical capacity, Sn-based anode materials for LIBs have many merits, such as low cost, environmental friendliness, and high energy density. SnO_2 converts into metallic Sn in the first cycle of the battery's operation, and subsequent Li storage happens via the reversible formation of Sn–Li alloys. However, its practical application as an anode material is significantly limited by the largely irreversible capacity loss that occurs during repeated charge/

discharge cycling. This is mainly caused by the large volume expansion/contraction (~300%) during Li+-ion insertion/extraction processes, which eventually leads to electrical disconnection arising from the crumbling and pulverization of the Sn domains [14, 16]. To circumvent this problem, metal oxides (e.g., SnO_2) and intermetallic compounds (e.g., SnSb) have been proposed. A variety of SnO_2 nanostructures, such as nanowires [12], nanosheets [52], hollow structures [87–89], nanocrystals [90], and porous structures [91, 92], as well as SnO_2-CNTs [93] or their hybrids [94], have been developed to improve the Li storage performance of SnO_2 anode materials. Although amorphous Li_2O produced during the initial cycles in SnO_2 or stepwise volume expansion in SnSb can accommodate the volume strain to some extent, in the composite electrodes, the carbon or its allotropes not only help to buffer the volume strain and maintenance of the structural integrity of Sn domains but also to improve the electrical conductivity of the composites. Nevertheless, it remains a challenge to reduce the irreversible capacity loss in the first cycle and increase the reversible capacity at a high rate by fabricating hierarchical SnO_2 architectures with deliberate high-order nanostructures. Most of these combinations have proven to be effective, revealing that the preparation of composites with 2D nanosheets of graphene is a promising strategy in developing high-energy anodes for LIBs. The introduction of graphene into Sn-based anode materials addresses the problem of large volume changes, resulting in high reversible capacity, rate capability, and structural stability of the composites. A large number of efforts have been reported to improve the cycling stability of Sn-based anodes with hierarchical SnO_2/graphene architectures. The morphologies and particle sizes of Sn composites with GNS have a critical influence on the electrochemical performance of the nanocomposite. A proper combination of morphology and design could promote this composite as a possible replacement for conventional graphite anodes. The extremely high flexibility of GNS can be the perfect supporting matrix or coating layer for Sn-based anode materials and could effectively be used to tackle the problem of the volume expansion of SnO_2 during the charge/discharge cycling process. The highly electronically conducting 2D nanosheets of graphene make a highly conducting hierarchical 2D network between the Sn active material in the system, which provides a medium for Li^+-ion/electron transfer during the charge/discharge process, and the beneficial synergistic effect between graphene and Sn active nanoparticles prevent the aggregation or crumbling and restacking of SnO_2 nanoparticles.

Zhao et al. induced vacancy into a metal oxide–graphene composite anode, which led to good electrochemical properties [66]. The composite anode showed a reversible capacity close to its theoretical value after five cycles, exhibiting a capacity loss of only 0.14% per cycle. The composite was able to retain 83% of its theoretical value after 150 cycles. High maintenance of the capacity was due to the shortened path for Li^+-ion diffusion in the electrode, resulting in fast lithiation/delithiation. The initial irreversible capacity loss was attributed to the reaction of Li^+-ion with traces of residual oxygen and hydrogen atoms in graphene and the formation of the SEI [66]. In fabricating a nanoporous composite anode of GNS and SnO_2, GNS were dispersed in ethylene glycol in presence of rutile SnO_2 nanoparticles [76]. Transmission electron microscope (TEM) analysis showed homogeneously distributed GNS loosely packed SnO_2 nanoparticles. The composite anode of GNS/SnO_2 showed a reversible capacity of 810 mAh g^{-1}, which is highly enhanced compared to pristine SnO_2 nanoparticles. The composite further exhibited a retention of 70% reversible capacity after 30 cycles. GNS acted as a dimension confinement of the SnO_2 nanoparticles, preventing volume expansion when Li^+-ions were inserted. Further, the pores developed between the GNS and SnO_2 was thought to act as a buffer space during charge/discharge cycles. SnO_2 and GO are dispersed in water in the presence of $SnCl_2.2H_2O$ as raw materials, resulting in a composite anode with a large current discharge capacity and high conductivity [95]. The initial charge/discharge capacities were 1923.5 and 1995.8 mAh g^{-1}

respectively. The reversible discharge capacity remained high even after 40 cycles at a current density of 1 A g^{-1}, showing the robustness of the composite anode for high-energy-storage LIBs.

6.3.2 Carbon-Coated SnO$_2$/Graphene Composite Anode

Sandwich-like hybrid nanosheets consisting of graphene-wrapped SnO/SnO$_2$ nanocrystals anchored on graphene (SnOx@graphene/graphene where x = 1 or 2) have been synthesized via an easy layer-by-layer self-assembly approach [96] or a single-phase co-precipitation method using N-methylpyrrolidone to exfoliate graphene from a graphite bar in the presence of cetyltrimethyl-ammonium bromide [97]. The GNS-SnO$_2$ composite prepared using the co-precipitation method delivers a capacity of ~1009 mA g^{-1} and retains 57% of the initial capacity after 10 cycles at a current density of 100 mA g^{-1}. Even at a high current density of 300 mA g^{-1}, the composite retains 81% of the initial capacity (~305.8 mA g^{-1}). The initial discharge/charge capacity of the hybrid nanocomposite of SnOx@graphene/graphene formed by layer-by-layer self-assembly are 1509 and 1007 mAh g^{-1}, respectively, with a coulombic efficiency of 66.7% at a current rate of 50 mA g^{-1}. The irreversible capacity in the first cycle is mainly attributed to the formation of Li$_2$O and SEI layers, which is confirmed in the CV studies. At a current density of 200 mA g^{-1}, the SnOx@graphene/graphene composite delivers a capacity of 1474 and 904 mAh g^{-1} in the initial discharge/charge process, benefiting from the unique double protection of graphene layers, while the SnO$_2$/graphene composite gives a capacity of 1140 and 230 mAh g^{-1} in the first discharge/charge process. Compared to the SnOx@graphene/graphene composite, the capacity of SnO$_2$/G decreases continuously due to the large volume change during cycling. The higher reversible capacity and cycling stability of the SnOx@graphene/graphene composite is attributed to the unique sandwich structure, which effectively provides an elastic buffer space to accommodate the volume changes during cycling and secures the SnOx nanocrystals between graphene layers. Also, the outermost graphene layers in the hierarchical structure could inhibit direct contact between the electrolyte and SnOx nanocrystals, which can prevent the side reaction with the electrolyte. SnO_2/GNS-based binary composites were further coated with amorphous carbon as a secondary buffering matrix to improve the cycling performance of Sn-based anodes for LIBs [5, 94, 98–104]. Zhang et al. reported achieving an exceptional electrochemical perfor-mance of graphene-SnO_2 electrodes by adding a thin-layer coat-ing of amorphous carbon on the nanosheets after SnO_2 decoration (GNS/SnO_2/C), using a hydrothermal technique and subsequent calcinations [102]. The GNS/SnO_2/C anode delivered a capacity of 1310 mAh g^{-1} in the first discharge and 958 mAh g^{-1} in the first charge process, while GNS/SnO_2 and SnO_2/C give 1423 and 1433 mAh g^{-1} in the first discharge and 978 and 902 mAh g^{-1} in the first charge process, respectively. The initial specific capacity of GNS/SnO_2/C is lower than that of GNS/SnO_2 due to the presence of carbon and graphene; however, it exhibits much better capac-ity retention than GNS/SnO_2. The high capacity of 757 mAh g^{-1} was achieved by the GNS/SnO_2/C composite at a current density of 200 mA g^{-1} even after 150 cycles with a coulombic efficiency of 100%, indicating superior reversibility. The superior cycling sta-bility of carbon-coated SnO_2/GNS composite electrodes is attrib-uted to the synergism between high conductivity and the large surface area of graphene, the effects of the coating layer of carbon, and the nanosized particles of SnO_2, which alleviate the effects of volume changes and help to stabilize the sandwich structure and/or conductivities of SnO_2 between GNS and the amorphous carbon coating layer. The GNS, properly wrapped over the SnO_2 nanoparticles, and the coated carbon layer work in synergy to effectively maintain the stability of the structural arrangement, avoid detachment of SnO_2 nanoparticles, serve as good electron conductors, and allow the easy access of Li^+-ions [105–107]. The surface carbon coating layer also effectively separates SnO_2/GNS, thereby preventing aggregation, which allows successful fabrica-tion of the carbon layer. Zhao et al. also reported similar effects on carbon-coated Li_2SnO_3/graphene composites [99] and a sandwiched hierarchical structure composed of a graphene substrate, intermediate SnO_2 nanorod arrays, and a carbon coating as an anode for Li storage [101]. The ternary composite exhibited a high reversible capacity and good cycling stability with high coulombic efficiency. A reversible capacity of ~736 mAh g^{-1} after 50 cycles at 60 mA g^{-1} current density has been reported [99]. In both cases, the carbon and anchored GNS buffered the large volume changes experienced by the anode during the insertion/extraction of Li^+-ions into the system. In the sandwiched structure, the carbon moieties also prevent the cracking and isolation of the SnO_2 nanorods during cycling due to the strong internal chemical bonding between the SnO_2 nanorods and the amorphous carbon layer. Later, Wang et al. [98] reported a novel hierarchical carbon-encapsulated-Sn (Sn@C) embedded GNS composite produced via a simple and scalable one-step CVD procedure [98]. In this novel hybrid system, the GNS-supported Sn@C core-shell structures consist of a crystalline Sn core, which is thoroughly covered by a carbon shell; extra care is taken to produce voids between the carbon shell and Sn core to increase the surface area and riley the large volume expansion of the Sn core during the charge/discharge cycling. The flexible and robust GNS acts as a template-like support, allowing the outer carbon shell to better confine the Sn nanoparticles and improve their mechanical and electronic properties. The hybrid Sn@C/GNS core-shell composite anode is reported to have a discharge capacity of 566 mAh g^{-1} after 100 cycles. Lian et al. reported a different approach to preparing SnO_2@C/GNS ternary nanocomposites [108], which combine hydrothermal self-assembly and thermal treatment at high temperature. The hybrid nanocomposites exhibit excellent cycling stability and rate capability due to the double conductive network formed by the amorphous conducting carbon coating anchored with a GNS buffering matrix.

Resolving volume expansion and other inherent issues plaguing the Sn/graphene-based composites, hybrid electrodes with other carbon materials, such as CNTs [3, 22, 23] and CNF [109], have been reported. Zou et al. reported a unique corn-like graphene/SnO₂/CNF hybrid electrode with a unique morphology, produced by growing SnO₂ nanoparticles on the surface of poly(vinylpyrrolidone) (PVP)-based CNF enveloped by graphene layers [109]. The graphene/SnO₂ electrode hybrid with CNF delivered a remarkable capacity of 1246.3 mAh g^{-1} at 0.5 A g^{-1}. Figure 6.2 schematically illustrates the electronic transport route and the volume effect of graphene/SnO₂/CNF during the charge/discharge process. The inner CNF on which the SnO₂ nanoparticles are anchored and the outer protecting GNS effectively hinder the volume expansion that occurs during Li⁺-ion insertion into SnO₂. In another report, a double protection strategy was employed to improve the electrode performance of SnO₂/graphene nanostructured electrodes introducing CNF. An electronically conducting 3D network is fabricated by simply mixing GNS with SnO₂ nanoparticles decorated on CNF prepared by electrospinning [109]. In this system, CNF and GNS facilitate the formation of a 3D electronic conductive network with SnO₂.

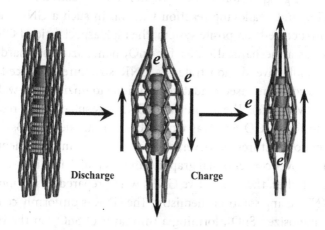

FIGURE 6.2 Schematic representation of the electronic transport route and the volume change during the Li⁺-ion insertion/extraction process in a GO/SnO₂/CNF anode. Reproduced from Zou et al. [109].

The homogeneously deposited 2D CNF increases the clearance between graphene layers, which provides a large number of active sites to adsorb Li^+-ions and allows for easier migration of Li^+-ions during the Li insertion/extraction process. The unique conductive 3D network formed by the CNF and GNS network ultimately enhances conductivity and shortens the electronic transport length within the electrode [109]. As a result of the synergetic action, the highly scattered SnOx@CNF@G composite exhibits a highly reversible capacity and excellent rate capability.

6.3.3 SnO_2/Graphene Nanoribbon Composite Anode

SnO_2/GNR is a structural variety designed to address the problem of SnO_2 aggregation and has been demonstrated to be a promising anode for high-energy LIBs [110–112]. This concept of developing novel nanostructures is broadly based on downsizing the dimensions of individual particles to nanosize, which can tolerate the strain associated with the volume expansion of active electrode materials much better by forming a breathable structure. Due to their extremely small size, these structures are capable of undergoing expansion and contraction during conversion or an alloying and dealloying reaction with Li. In such a GNR-based nanostructured composite system, the high aspect ratio of GNRs serves as a mechanical buffer for SnO_2 nanoparticles. Bhardwaj et al. [50] have demonstrated a GNR structure produced by using a solution-based oxidative technique to unzip multi-walled CNTs (MWNTs). The as-prepared GNS was subjected to reduction to form rGO and was explored as an anode in LIBs [50]. The nanostructures, however, showed minimal improvement in capacity over conventional graphite anodes [50]. Later, in two different studies, the conductive GNRs were prepared by unzipping MWNTs using solution chemistry. The GNR is uniformly coated with nanosized SnO_2, forming a thin layer of SnO_2 on the rGO surface. The resulting SnO_2/rGO nanoribbon anode materials demonstrated high capacity, good rate performance, and excellent cycling operation. A first discharge/charge capacity of >1500 mAh g^{-1} [110] was reported. A reversible discharge capacity of 1027 mAh g^{-1} at 0.1 A g^{-1} after 165 cycles and 640 mAh g^{-1} at 3.0 A g^{-1} after 160 cycles was observed. At 1.0 A g^{-1} current density, no capacity decay was found even after 600 cycles with respect to that at second cycle [111]. It was also shown that the reversible capacity retains around 825 mAh g^{-1} at a current density of 100 mA g^{-1} with a coulombic efficiency of 98% after 50 cycles [110]. Table 6.1 summarizes the electrochemical properties of Sn/graphene composite electrodes in LIBs. The high reversible capacity, good rate performance, and excellent cycling stability of the composite are attributed to the synergistic combination of electrically conductive GNR networks with large aspect ratios as well as the homogeneous distribution of small-sized SnO_2 nanoparticles along or between GNR stacks. The conductive GNR network may also facilitate electrolyte diffusion and Li$^+$-ion transport into the nanosized SnO_2 and/or act as conductive additives that buffer the volume expansion and contraction of SnO_2 during Li insertion/extraction.

6.3.4 SnO_2/Graphene/Fe_2O_3 Ternary Composite Anode

In addition to various carbon materials, different electrochemically active metallic oxides that exhibit low volume change have been used to prepare Sn-based ternary hybrid electrodes for LIBs [113–115]. Fe_2O_3 is considered one of the most practically attractive anode materials that react with Li via conversion reaction mechanisms. Similar to SnO_2, Fe_2O_3 electrode materials suffer from significant volume change upon reaction with Li, resulting in pulverization and cracking of electrodes in the battery. Volume expansion for an Fe_2O_3 electrode is about 96% [116], less than that of SnO_2 (300%) [117] during lithiation/delithiation. To accommodate the high level of volume alteration of SnO_2 without damaging the structure of the electrode, novel nanostructured electrode architectures of SnO_2/Fe_2O_3–graphene were developed to tackle the volume expansion and related cracking problems of these electrodes. The use of Fe_2O_3/SnO_2/graphene ternary

TABLE 6.1 Electrochemical Performance of Various Sn/Graphene-Based Binary LIB Electrodes

Materials	Feature	Electrochemical Performance			
		Current Density	Cycle No.	Capacity Retention (mAh g^{-1})	Ref.
SnO$_2$/graphene composites	SnO$_2$ nanoparticles with an average diameter of 3 nm	2 A g^{-1}	1000	1813	[143]
SnS$_2$/GNS nanocomposites	Thicknesses of the SnS$_2$ plate and the graphene sheet are measured to be 25 nm and 12 nm	1.6 mA cm^2	100	704	[144]
SnS$_2$/RGO nanocomposites	Plentiful SnS$_2$ nanocrystals deposited on flexible RGO	0.1 C	200	1034	[145]
Co$_2$SnO$_4$ hollow cubes@RGO nanocomposites	Co$_2$SnO$_4$ hollow cubes encapsulated in graphene	100 mA g^{-1}	100	1000	[146]
Hollow Zn$_2$SnO$_4$@graphene composites	Hollow boxes supported by flexible graphene sheets	300 mA g^{-1}	45	752	[99]
Graphene networks anchored with Sn@graphene	3D porous graphene networks anchored with Sn nanoparticles encapsulated in graphene shells	200 mA g^{-1}	340	1076	[147]
Graphene nanoribbons (GNRs) and SnO$_2$	GNRs prepared using sodium/potassium unzipping of MWNT	100 mA g^{-1}	50	825	[110]

nanocomposites as anodes for LIBs have been reported by many researchers [113, 114, 118]. In two different studies, Xia et al. [114] synthesized the ternary nanocomposite of SnO_2/Fe_2O_3/graphene via homogeneous in-situ precipitation of Fe_2O_3 nanoparticles onto graphene oxide followed by reduction of graphene oxide with $SnCl_2$ [114], and Liu et al. [113] prepared the hybrid anode via an easy hydrothermal route [113]. The core-shell structured $Fe_2O_3@SnO_2$ nanoparticles were first synthesized using a hydrothermal method, and the resultant spindle-like $Fe_2O_3@SnO_2$ nanoparticles were coated on a flexible graphene oxide film through vacuum filtration, followed by thermal reduction [113]. In both cases, the hybrid nanocomposite showed superior electrochemical performance, cycling stability, and rate capability. The novel ternary nanocomposite exhibits an initial discharge/charge capacity of 1179 and 746 mAh g^{-1}, respectively, at a high current density of 400 mA g^{-1} with a coulombic efficiency of ~63%. However, after a few cycles of the charge/discharge cycle, the coulombic efficiency increased by >90% and maintained a specific capacity above 700 mAh g^{-1} for 100 cycles. To understand the effect of Fe_2O_3 in the ternary nanocomposites, a comparison study was carried out on electrochemical performance with the rGO/SnO and rGO electrodes. In the absence of Fe_2O_3, the rGO/SnO_2 anode shows mediocre performance with a specific capacity of 600 mAh g^{-1} at the first discharge/charge cycle, followed by approximately 30% capacity loss after only 100 cycles at a current density 400 mA g^{-1}, while the rGO anode exhibits a reversible capacity of 550 mAh g^{-1} after 100 cycles, indicating a capacity loss of 18%. The rGO/SnO_2/Fe_2O_3 hybrid electrode is reported to have good rate capability. It shows a reversible capacity of 623 and 139 mAhg^{-1} at 0.5 C, and 10 C respectively. A reversible capacity of 782 mAh g^{-1} was recovered after decreasing the charge rate from 10 C to 0.2 C, indicating that the hybrid electrode had excellent capacity reversibility and cycle stability [114]. The decreased specific capacity at a higher current rate is due to the increasing resistance at the interface caused by the formation of SEI between the electrolyte and the active materials as well as insufficient Li^+-ion diffusivity at high rates [119, 120]. Zhu et al. [121] also prepared ternary-phased SnO_2-Fe_2O_3/rGO (termed SnO_2/Fe_2O_3/rGO) composite nanostructures, which delivered a remarkable capacity of 958 mAh g^{-1} at 395 mA g^{-1} [121]. Similar results have also been demonstrated by Lian and co-workers [122], where the Fe_3O_4/SnO_2/graphene ternary nanocomposite delivered a higher reversible capacity of 1198 mAh g^{-1} at 100 mA g^{-1} [122]. In all the studies, the enhanced cycling stability and specific capacity of the rGO/Fe_2O_3/SnO_2 anode are attributed to the complementary roles among Fe_2O_3, SnO_2, and rGO in the nanostructures. The flexible graphene component in the hybrid system forms a 2D electronically conducting network which facilitates the easy transfer of Li^+-ions and electrons and rapid electrolyte diffusion channel. The GNS also serves as an effective elastic buffer to relieve the stress that would otherwise accumulate in the Fe_2O_3/SnO_2 particles during the charge/discharge process, thereby aiding the Fe_2O_3/SnO_2 in retarding the electrochemical aggregation of nanoparticles. As in the binary composite, in the ternary hybrid nanocomposite, the dispersion of Fe_2O_3/SnO_2 nanoparticles on graphene mitigates the degree of stacking of the graphene sheets. In addition, the SnO_2 nanoparticles in the hybrid composite play a crucial role in hindering the agglomeration of Fe_2O_3 nanoparticles or vice versa during the conversion reaction or alloying or dealloying reaction during the lithiation/delithiation process.

6.3.5 SnO_2/Graphene/Polymer Ternary Composite Anode

Some novel ternary SnO_2/graphene/polymer based nanocomposites anodes have been developed to enhance the performance of LIBs [18, 19, 50, 93, 123–126]. Each study reported on structures composed of graphene/SnO_2/polymer or rGO/SnO_2/polymer nanocomposite produced by via one-pot synthesis, resulting in a thin conducting polymer film on the surface of the graphene/SnO_2 structure. These structures were reported to show good electrochemical performance when used as anode/electrode materials. The conducting polymer and GNS serve as buffering matrices, which relieve the stress caused by the volume expansion of Sn-based anodes during the lithiation/delithiation process, resulting in better cycling and rate capabilities in the LIBs. Conducting polymers, such as polyaniline (PANI), polypyrrole (PPy) and poly(3,4-ethylenedioxythiophene) (PEDOT), provide a conductive backbone for the active materials, and their soft structured matrix can buffer the internal stress of electrodes suffering from large volume changes [127, 128]. The intrinsically conducting polymer thus offers a good tradeoff between mechanical stability and electrical conductivity for the active electrode [129].

Zhao et al. [130] reported a new ternary graphene/Li_2SnO_3/PPy composite [130] in which the co-introduction of PPy and graphene created a double buffering structure for Li_2SnO_3 electrodes, which resulted in a reversible capacity of ~700 mAh g^{-1} after 30 cycles. Hollow spheres of SnO_2/GO/PEDOT were synthesized by imbedding them in GO and enveloping them in a sheath of PEDOT [19]. The hollow structures of SnO_2 can partially accommodate the large volume change, delaying capacity fading and imparting efficient catalytic activity and structural stability [130]. The study adopted the hydrothermal synthesis of SnO_2 hollow spheres using an organic bifunctional molecule followed by the formation of a hybrid electrode with GNS and PEDOT. While GNS serves as a scaffold for entrapment of SnO_2 hollow spheres and prevents their coalescence during electrochemical cycling, the PEDOT overlayer, being a robust conducting polymer, effectively inhibits the disintegration of the SnO_2 hollow sphere during charge/discharge cycling, and cumulatively GNS and PEDOT afford capacity retention with cycling [19].

The optimized SnO_2 hollow sphere displayed a reversible Li$^+$-ion storage capacity of 400 mAh g^{-1} attained at a current density of 100 mA g^{-1} after 30 charge/discharge cycles. The key role of GNS is to form the electron-conducting 2D network for the SnO_2 hollow sphere and that of PEDOT is to form the polymer envelope, which buffers the severe volume changes that occur during the lithiation/delithiation process in cycling; this inhibits electrode disintegration and rapid capacity fading, thereby amplifying the reversible capacity, rate capability, and cycling stability of the hybrid electrode. GNS forms a highly conducting network for Li^+-ion/electron transfer during charge/discharge process, alleviating the mechanical stress caused by the severe volume changes during Li insertion/extraction and reducing the detachment of electroactive material from the electrode. The buffering GNS and PEDOT layer also inhibit the aggregation of SnO_2 nanoparticles. Conductive polymers, such as PANI, PEDOT, and PPy, can prevent direct contact between Sn-based crystallites and the electrolyte and thus inhibit any side chemical reactions of the electrolyte with the oxide. However, the large Li storage capacity can be linked to the microstructure and high surface area of the hollow spheres, wherein the porous shells allow the innards of the sphere to be electrolyte accessible. For comparison, the cycling properties of SnO_2 solid microspheres have been studied, showing that the solid microspheres exhibit an initial reversible capacity of 716 mAh g^{-1} and that after 30 cycles, the capacity faded to 141 mAh g^{-1}. The very high capacity of the SnO_2 hollow sphere compared to solid microspheres is mainly because of the open porous structure of the SnO_2 hollow spheres, which allows greater ion uptake during electrochemical cycling.

The exceptionally improved cycling performance of the SnO_2 hollow spheres compared to solid microspheres is due to the relative ease with which the SnO_2 hollow spheres can accommodate volume expansion upon Li insertion/extraction. Enveloping the SnO_2 hollow spheres embedded in GNS resulted in a significant improvement in cycling performance and excellent electrochemistry. The SnO_2 hollow sphere/GNS/PEDOT hybrid anode showed initial discharge/charge capacities of ~1654 and ~717 mAh g^{-1} respectively at 100 mA g^{-1} current density with a coulombic efficiency of 43.4% when the cell was cycled between 0.01–2.5 V. The large irreversible capacity loss was mainly because of the irreversible conversion reaction of SnO_2 to Sn and SEI layer formation, which is unavoidable for most metal oxides and composites with carbonaceous moieties. When the mass of only the SnO_2 hollow spheres in the hybrid was considered, the electrode showed an initial reversible capacity of ~1468 mAh g^{-1} and ~1248 mAh g^{-1} after 150 cycles at a current density of 100 mA g^{-1}. The coulombic efficiency of the ternary hybrid for the third cycle reached 96.5% and gradually increased with cycling. After 150 cycles, the hybrid electrode showed a reversible capacity of ~609 mAh g^{-1} with a coulombic efficiency of more than 99%. The SnO_2 hollow spheres/GO/PEDOT hybrid also performed well in terms of rate capability as well as capacity and showed a capacity of 381 mAh g^{-1} at extremely high current density of 2000 mA g^{-1}, which in turn demonstrated the viability of the SnO_2 hollow spheres/ GO/PEDOT composite hybrid as a high-performance anode for LIBs [19].

Ding et al. [93] reported the use of a 3D nanostructure composed of ternary SnO_2/polyaniline/graphene (SnO_2@PANI/rGO) nanohybrids as anode materials by a simple dip-coating on Cu foam. In the 3D nanostructures, PANI acts as the conductive matrix as well as a good binding agent for SnO_2 nanoparticles and GNS, greatly improving the rate performance. The as-prepared SnO_2@PANI/rGO showed good electrochemical performance with initial discharge/charge capacities of 1468 and 877 mAh g^{-1} respectively, with an initial coulombic efficiency of 60% (initial reversible capacities ~772 mAh g^{-1} and became stable around ~750 mAh g^{-1}) after 100 cycles at a current density of 100 mA g^{-1}, which is significantly higher than the binary composite SnO_2/rGO (458 mAh g^{-1}), SnO_2@PANI (480 mAh g^{-1}), and pristine SnO_2 nanoparticles (300 mAh g^{-1}). Coulombic efficiency is gradually increased to reach 97% after 10 cycles, which in turn increased up to ~100% after 100 cycles. The ternary nanohybrids also exhibit excellent rate capability. The hybrid anode delivers a discharge capacity of 268 mAh g^{-1} at 1000 mA g^{-1}, and a high specific capacity of ~748 mAh g^{-1} can be recovered when the current density returns back to 100 mA g^{-1}, which clearly indicates good cycling stability. More significantly, after 100 cycles at 100 mA g^{-1}, the ternary nanohybrids maintained a high specific capacity of 749 mAh g^{-1}, which is 269–449 mAh g^{-1} higher than SnO$_2$/rGO, SnO$_2$@PANI, and pure SnO$_2$ nanoparticles [93, 123].

6.3.6 SnO$_2$/Graphene/TiO$_2$ Ternary Composite Anode

TiO$_2$ and TiO$_2$-based composite materials have been widely prepared and applied in LIBs due to their long cycle life, small volume change, lower self-discharge rate, high safety during Li insertion/extraction, low cost, and environmentally benign nature [131–135]. Generally, these metal oxides insert or release the Li$^+$-ions mainly via alloying or conversion reaction [136, 137]. Anatase TiO$_2$, however, can accommodate Li into the vacant sites of the crystalline structure, which can be viewed as a stacking of zigzag chains consisting of highly distorted edge-sharing TiO$_6$ octahedra [138, 139] corresponding to a theoretical capacity of 167.5 mAh g^{-1} [140, 141]. Because of this unique structure, TiO$_2$ shows a negligible volume change (3%–4%) [139] and can be used as the stable barrier for SnO$_2$-based anodes for LIBs. The effect of TiO$_2$ on the charge/discharge and rate performance of SnO$_2$/TiO$_2$/GNS ternary nanocomposites have been widely reported [115, 142]. Jiang et al. [115] anchored the in-situ–formed SnO$_2$ and TiO$_2$ nanoparticles on the surface of flexible rGO nanosheets by a solvothermal method combined with a hydrothermal two-step method [115]. Electrochemical studies showed that during the Li$^+$-ion insertion process (discharging), the SnO$_2$/TiO$_2$/GNS electrode delivered a first discharge capacity of 1952 mAh g^{-1}, while during the subsequent Li$^+$-ion extraction process (charging), it gave only a capacity of 954 mAh g^{-1} at a current density of 50 mA g^{-1} (coulombic efficiency of 49%). Compared to an SnO$_2$/TiO$_2$/GNS electrode, the binary composite of the SnO$_2$/GNS electrode delivered a larger discharge capacity of 2805 mAh g^{-1} and a charge capacity of 1504 mAh g^{-1} in the first cycle (coulombic efficiency of 54%). The higher initial capacity of the SnO$_2$/GNS electrode arises mainly from the significantly larger difference in the SnO$_2$ content in SnO$_2$/GNS nanocomposites (about 90%)

than that in $SnO_2/TiO_2/GNS$ nanocomposites (about 81%), which gives a high theoretical capacity of 782 mAh g^{-1} compared to the TiO_2 active material. The specific capacity of the SnO_2/GNS electrode is higher up to the sixth cycle, while the $SnO_2/TiO_2/GNS$ electrode showed superior cycling performance over extended cycling. For the SnO_2/GNS anode, the reversible capacity continuously decreased up to the first 30 cycles and reached to 332 mAh g^{-1} after 50 cycles, while that of the $SnO_2/TiO_2/GNS$ anode tended to level off after 15 cycles and showed a reversible capacity of 537 mAh g^{-1} during the 50th cycle. The major reason behind the poor cyclability of the SnO_2/GNS anode was the drastic volume change in electrode material caused by the agglomeration of SnO_2 nanoparticles during the charge/discharge cycling process, which led to electrode fracture and discontinuity/cracking of the electrode. In addition, the formation of an unstable SEI film results in poor electronic conductivity, which is also a major reason for the fast decaying of specific capacity over cycling. The coulombic efficiency of the $SnO_2/TiO_2/GNS$ electrode changed from 49% in the first cycle to over 97% in the subsequent cycles, which is due to the presence of anatase TiO_2 nanoparticles in the hybrid anode. The small amount of TiO_2 nanoparticles in the hybrid electrode could act as stable barriers to prevent the agglomeration of SnO_2 nanoparticles on the flexible rGO sheets and could serve as clearance space to accommodate the volume change during the charge/discharge process, maintaining the stability of the electrode structure and avoiding the formation of an unstable SEI resistive layer. The as-synthesized ternary $SnO_2/TiO_2/GNS$ nanocomposites also showed good rate capability, with a reversible capacity of 250 mAh g^{-1} even at a current density of 1000 mA g^{-1}.

Based on the above discussion, it can be concluded that introducing double buffering matrices such as ultrathin carbon coating, anchoring the active electrode material with CNF or CNTs, or mixing SnO_2 with conducting polymer combined with GNS creates an electronically conducting 3D carbon network. This could enhance the electronic conductivity of the electrode and

effectively suppress the aggregation of nanoparticles and unde-
sired side reactions with electrolyte, thereby reducing or sta-
bilizing the SEI layer, resulting a drastic enhancement in the
electrochemical performance of Sn-based anodes.

6.4 SUMMARY

Since the historical discovery of graphene, LIBs and their related
technologies have moved to the forefront of the technology
field, becoming an inevitably important part of day-to-day life.
According to Moore's law, the number of transistors in densely
integrated circuits doubles every year while the cost is halved, and
LIB technology is struggling to keep up. While LIBs are becoming increasingly important to human life for use in portable
electronics as well as for sustainable energy for green transportation, there is still plenty of room at the bottom for improvement. The flurry of ongoing research into novel anodic materials
or composites has found graphene to be a potential material for
the development of high-performance LIBs. As it is the thinnest,
strongest, and most highly conductive of modern materials, it has
the potential to revolutionize LIB technology. Compared to pure
carbon-based anode materials, graphene has a higher Li storage
capacity (theoretical capacity 740 mAh g^{-1}); however, its specific
capacity is much less compared to other popular Li storage compounds such as metal, metal oxides (SnO_2), and other inorganic
materials like silicon (4200 mAh g^{-1}). These compounds, however,
suffer from problems with relatively low electronic conductivity
or large volume changes, which limit their practical applications
in LIBs. To address this problem, graphene has been extensively
investigated as a conducting additive to enhance electronic conductivity and minimize volume changes during continuous
charge/discharge cycling. GNS with various nanoparticles have
been reported as anode materials for LIBs, not only enhancing
electrochemical performance but also resulting in improved
functionality and properties due to the interaction between the
GNS and active materials. GNS can buffer the volume changes of

active materials and prevent them from pulverization. Among the different anodic materials for LIBs, SnO_2 is considered to be the most promising due to its high theoretical specific capacity (782 mAh g^{-1}) and low potential for Li alloying. As SnO_2/GNS composites are thus widely studied as an anode material, this chapter has focused on the effect of GNS on the electrochemical properties of SnO_2 anodes and different approaches and fabrication methods.

REFERENCES

1. Reddy et al. *Handb. Batter.* (2002) 34.1–34.62.
2. Yazami et al. *J. Power Sources* 9 (1983) 365.
3. Chen et al. *Electrochim. Acta* 108 (2013) 674.
4. Bondarenko et al. *Carbon N. Y.* 36 (1998) 1107.
5. Cheng et al. *J. Power Sources* 232 (2013) 152.
6. Ishikawa et al. *J. Power Sources* 62 (1996) 229.
7. Wan et al. *Chem. Commun.* 51 (2015) 9817.
8. Peled et al. *J. Electrochem. Soc.* 143 (1996) L4.
9. Zhao et al. *Phys. Scr.* 2010 (2010) 014036.
10. Besenhard. *Carbon N. Y.* 14 (1976) 111.
11. Besenhard et al. *J. Electroanal. Chem. Interfacial Electrochem.* 53 (1974) 329.
12. Park et al. *Angew. Chemie Int. Ed.* 46 (2007) 750.
13. Cao et al. *J. Mater. Chem.* 22 (2012) 9759.
14. Wang et al. *Adv. Mater.* 24 (2012) 1903.
15. Wu et al. *Mater. Lett.* 107 (2013) 27.
16. Deng et al. *Energy Environ. Sci.* 2 (2009) 818.
17. Mutoro et al. *Energy Environ. Sci.* 4 (2011) 3689.
18. Wang et al. *RSC Adv.* 2 (2012) 10268.
19. Bhaskar et al. *J. Phys. Chem. C* 118 (2014) 7296.
20. Wang et al. *J. Phys. Chem. C* 115 (2011) 11302.
21. Qu et al. *Mater. Lett.* 59 (2005) 4034.
22. Li et al. *J. Mater. Chem. A* 2 (2014) 2526.
23. Chen et al. *RSC Adv.* 2 (2012) 11719.
24. Lu et al. *Int. J. Electrochem. Sci.* 7 (2012) 6180.
25. Veeraraghavan et al. *J. Electrochem. Soc.* 149 (2002) A675.
26. Xu et al. *JPH Chem.* 3 (2012) 309.
27. Zhu et al. *Small* 10 (2014) 1.
28. León et al. *ACS Nano* 8 (2014) 563.

29. Liang et al. *Anal. Chem.* 63 (1991) 423.
30. Yumin et al. *J. Korean Phys. Soc.* 58 (2011) 938.
31. Benwadih et al. *Mater. Appl.* 15 (2014) 614.
32. Ervin et al. *Electrochim. Acta* 147 (2014) 610.
33. Lee et al. *Chem. Eng. J.* 230 (2013) 296.
34. Gupta et al. *Prog. Mater. Sci.* 73 (2015) 44.
35. Zhang et al. *J. Mater. Chem.* 21 (2011) 1673.
36. Xiang et al. *Carbon N. Y.* 49 (2011) 1787.
37. Vargas et al. *Phys. Chem. Chem. Phys.* 15 (2013) 20444.
38. Mahmoud et al. *Desalination* 356 (2014) 208.
39. Yu et al. *Sci. Total Environ.* 502 (2015) 70.
40. Pham et al. *J. Mater. Chem.* 21 (2011) 3371.
41. Schniepp et al. *J. Phys. Chem. B* 110 (2006) 8535.
42. Xiang et al. *RSC Adv.* 2 (2012) 6792.
43. Wang et al. *Carbon N. Y.* 47 (2009) 2049.
44. Lian et al. *Electrochim. Acta* 55 (2010) 3909.
45. Guo et al. *Electrochem. Commun.* 11 (2009) 1320.
46. Yao et al. *Electrochem. Commun.* 11 (2009) 1849.
47. Lu et al. *Sci. Rep.* 4 (2014) 4629.
48. Wu et al. *ACS Nano* 5 (2011) 5463.
49. Cai et al. *Electrochim. Acta* 90 (2013) 492.
50. Bhardwaj et al. *J. Am. Chem. Soc.* 132 (2010) 12556.
51. Leela et al. *ACS Nano* 4 (2010) 6337.
52. Wang et al. *J. Mater. Chem.* 21 (2011) 5430.
53. Li et al. *J. Chem. Phys.* 129 (2008) 104703.
54. Mao et al. *Energy Environ. Sci.* 5 (2012) 7950.
55. Zheng et al. *Nat. Commun.* 5 (2014) 5261.
56. Xing et al. *Sci. Rep.* 6 (2016) 26146.
57. Wang et al. *Adv. Energy Mater.* 6 (2016) 1502100.
58. Liu et al. *J. Power Sources* 342 (2017) 157.
59. Tang et al. *Adv. Funct. Mater.* 19 (2009) 2782.
60. Zhao et al. *ACS Nano* 5 (2011) 8739.
61. Liu et al. *Adv. Mater.* 24 (2012) 1089.
62. Hu et al. *J. Power Sources* 237 (2013) 41.
63. Hu et al. *Electrochim. Acta* 91 (2013) 227.
64. Oh et al. *Electrochim. Acta* 135 (2014) 478.
65. Lu et al. *J. Mater. Chem. A* 2 (2014) 1802.
66. Zhao et al. *Adv. Energy Mater.* 1 (2011) 1079.
67. Banhart et al. *ACS Nano* 5 (2011) 26.
68. Lahiri et al. *Nat. Nanotechnol.* 5 (2010) 326.
69. Liu et al. *J. Nanomater.* 2013 (2013) 1.

70. Wang et al. *ACS Nano* 7 (2013) 2422.
71. Ji et al. *Energy Environ. Sci.* 4 (2011) 2682.
72. Pollak et al. *Nano Lett.* 10 (2010) 3386.
73. Valota et al. *ACS Nano* 5 (2011) 8809.
74. Uthaisar et al. *Nano Lett.* 10 (2010) 2838.
75. Kubota et al. *J. Phys. Soc. Japan* 79 (2010) 1.
76. Paek et al. *Nano Lett.* 9 (2009) 72.
77. Shi et al. *Electrochim. Acta* 246 (2017) 1104.
78. Hoshide et al. *Nano Lett.* 17 (2017) 3543.
79. Jiang et al. *ACS Nano* 11 (2017) 5140.
80. Hao et al. *Electrochim. Acta* 260 (2018) 965.
81. Zhou et al. *Ionics* 24 (2018) 1321.
82. Qian et al. *J. Electrochem. Soc.* 159 (2012) 1135.
83. Li et al. *Adv. Sci.* 2 (2015) 1500284.
84. Hwang et al. *Nano Lett.* 11 (2011) 4826.
85. Winter et al. *Electrochim. Acta* 45 (1999) 31.
86. Prabakar et al. *J. Power Sources* 209 (2012) 57.
87. Lou et al. *Adv. Mater.* 18 (2006) 2325.
88. Ye et al. *Small* 6 (2010) 296.
89. Wu et al. *Nanoscale* 10 (2018) 11460.
90. Liu et al. *ACS Appl. Mater. Interfaces* 10 (2018) 2515.
91. Han et al. *CrystEngComm* 13 (2011) 3506.
92. Kim et al. *J. Mater. Chem.* 18 (2008) 771.
93. Ding et al. *Adv. Funct. Mater.* 21 (2011) 4120.
94. Liang et al. *J. Solid State Chem.* 184 (2011) 1400.
95. Guo et al. *J. Power Sources* 240 (2013) 149.
96. Huang et al. *J. Mater. Chem. A* 2 (2014) 6249.
97. Ravikumar et al. *Phys. Chem. Chem. Phys.* 15 (2013) 3712.
98. Wang et al. *Phys. Chem. Chem. Phys.* 15 (2013) 3535.
99. Zhao et al. *Ceram. Int.* 39 (2013) 1741.
100. Su et al. *ACS Nano* 6 (2012) 8349.
101. Wang et al. *Energy Environ. Sci.* 6 (2013) 2900.
102. Zhang et al. *Carbon N. Y.* 50 (2012) 1897.
103. Zhang et al. *J. Mater. Chem. A* 5 (2017) 19136.
104. Selim et al. *Electrochim. Acta* 224 (2017) 201.
105. Morishita et al. *J. Power Sources* 160 (2006) 638.
106. Fan et al. *Adv. Mater.* 16 (2004) 1432.
107. Zhang et al. *Adv. Mater.* 20 (2008) 1160.
108. Lian et al. *Electrochim. Acta* 116 (2014) 103.
109. Zou et al. *J. Mater. Chem. A* 2 (2014) 4524.
110. Lin et al. *ACS Nano* 7 (2013) 6001.

111. Li et al. *Nano Res.* 7 (2014) 1319.
112. Li et al. *J. Alloy. Compd.* 729 (2017) 1064.
113. Liu et al. *J. Mater. Chem.* 2 (2014) 4580.
114. Xia et al. *ACS Appl. Mater. Interfaces* 5 (2013) 8607.
115. Jiang et al. *New J. Chem.* 37 (2013) 3671.
116. Lin et al. *J. Phys. Chem. Lett.* 2 (2011) 2885.
117. Wang et al. *Nano Lett.* 11 (2011) 1874.
118. Gu et al. *Nano Res.* 10 (2017) 121.
119. Kim et al. *J. Mater. Chem.* 22 (2012) 15514.
120. Han et al. *Chem. Commun.* 48 (2012) 9840.
121. Zhu et al. *J. Mater. Chem.* 21 (2011) 12770.
122. Lian et al. *Electrochim. Acta* 58 (2011) 81.
123. Ding et al. *Electrochim. Acta* 157 (2015) 205.
124. Liu et al. *Solid State Ionics* 294 (2016) 4.
125. Zhou et al. *ACS Appl. Mater. Interface* 8 (2016) 13410.
126. Liu et al. *RSC Adv.* 6 (2016) 9402.
127. Kim et al. *Adv. Energy Mater.* 3 (2013) 1004.
128. Han et al. *Adv. Funct. Mater.* 23 (2013) 1692.
129. Manuel et al. *Mater. Res. Bull.* 45 (2010) 265.
130. Zhao et al. *Ceram. Int.* 39 (2013) 6861.
131. Shin et al. *Adv. Funct. Mater.* 21 (2011) 3464.
132. Yang et al. *Adv. Mater.* 23 (2011) 3575.
133. Ali et al. *J. Mater. Chem.* 22 (2012) 17625.
134. Chen et al. *Chem. Rev.* 107 (2007) 2891.
135. Wang et al. *ACS Nano* 3 (2009) 907.
136. Wang et al. *Chem. Commun.* 47 (2011) 8061.
137. Cheng et al. *Adv. Mater.* 23 (2011) 1695.
138. Shubha et al. *Electrochim. Acta* 125 (2014) 362.
139. Nuspl et al. *J. Mater. Chem.* 7 (1997) 2529.
140. Chen et al. *Mater. Today* 15 (2012) 246.
141. Aravindan et al. *J. Mater. Chem. A* 1 (2013) 308.
142. Liang et al. *J. Alloy. Compd.* 673 (2016) 144.
143. Chen et al. *J. Mater. Chem. A* 2 (2014) 5688.
144. Liu et al. *ACS Appl. Mater. Interfaces* 5 (2013) 1588.
145. Mei et al. *J. Mater. Chem. A* 1 (2013) 8658.
146. Zhang et al. *J. Mater. Chem. A* 2 (2014) 2728.
147. Qin et al. *ACS Nano* 8 (2014) 1728.

CHAPTER 7

Graphene and Its Composites as Cathodes for Lithium Ion Batteries

7.1 INTRODUCTION

The wide range of cathode materials currently used for lithium ion batteries (LIBs) has proved to create a bottleneck in achieving high efficiency due to problems associated with them [1–3], for example, low energy density and power density, poor electrical conductivity, the formation of an solid electrolyte interface (SEI) layer, the agglomeration of particles generated from their nanostructures, and the sluggish kinetics of Li+-ion transportation [4]. A lot of effort, therefore, has been directed to improve on some of these problems inherent in current materials for the cathode in order to improve the overall performance of the battery. There have been attempts to use pure graphene [2] or graphene oxide (GO) [5] as cathode materials, but the majority of the research has involved the use of composite materials [6–9]. In composites, traditional materials are combined with graphene or its derivatives to make the cathode [4]. The composite cathode materials of graphene come in different "flavors," for example, $LiMPO_4$ (M = Fe, Co, Mn, Ni, V)-graphene composites [6, 7], Li metal (Mn, Co, Ni) oxide–graphene composites [10], sulfur–graphene composites, Li metal (Mn, Co, Ni) oxide–graphene composites, sulfur–graphene composites [11, 12], metal oxide–graphene composites [13], and other graphene-based composites [9]. Among the cathode materials produced, $LiFePO_4$, $LiCoO_2$, $LiMn_2O_4$, and V_2O_5 are the most interesting cathodes for LIBs.

7.2 LIFEPO$_4$/GRAPHENE COMPOSITE AS CATHODE MATERIAL

$LiFePO_4$ with its olivine structure is one of the most promising cathode materials for LIBs, owing to its high theoretical capacity (170 mAh g^{-1}), flat operating voltage (3.4 V vs. Li$^+$/Li), good cycling stability, abundance in availability, environmental friendliness, good thermal stability, and low cost [14]. Compared to commercial cathode materials, for example, $LiMPO_4$ (M = Fe, Co, Mn, Ni, V), the biggest advantage of $LiFePO_4$ is that it is non-toxic [15, 16]. Because of these advantages, $LiFePO_4$ has been the most extensively investigated [4] material for the cathode in LIBs. However, its intrinsically poor electronic conductivity (about 10^{-9}–10^{-10} S cm^{-1}) and low Li$^+$ transport capability (approximately 10–14 cm^2 s^{-1}) [17, 18] have been the major hurdle in its commercial use as a cathode in LIBs [9].

Many efforts have been made to improve the electronic conductivity, cycling stability, rate capability, and Li diffusion of $LiFePO_4$ by mixing the $LiFePO_4$ particle with conducting carbon, using nanoparticles with high surface area [19], doping with other metal ions [20], coating the $LiFePO_4$ particles with a conductive carbon layer [21], and aliovalent doping [22–24]. It has been demonstrated that metal ion doping not only expands the Li diffusion channel and reduces interface resistance, but also increases the output voltage of $LiFePO_4$. Reducing the size and increasing the surface area of the $LiFePO_4$ particle shortens the Li^+ diffusion path and improves the wettability of the electrode with the electrolyte; however, it leads to an interface effect, which adversely affects electrochemical performance [25]. A carbon coating has been successfully used to improve the electrical conductivity and cycling stability of $LiFePO_4$, though it may lower the energy density of the LIBs [25]. Additionally, a carbon coating can be easily accomplished through an in-situ pyrolysis of organic carbon precursors, such as sucrose [27], glucose [28], starch [29], citric acid [30], ascorbic acid [31], adipic acid [32], pitch carbon [27], polypropylene [31], poly(vinylpyrrolidone) (PVP) [33], polyvinyl alcohol [34], polythiophene [35], and polyacene [36] on $LiFePO_4$. However, the composition, graphitization extent, thickness, surface functionality, and uniformity of the carbon coating layer are difficult to control in practice, which significantly affects the electrochemical performance of LIBs in practice [37, 38].

Recently, a composite of $LiFePO_4$ with graphene and its nanostructures has garnered much attention for its ability to improve the electrochemical performance of $LiFePO_4$-based cathodes. The crystalline $LiFePO_4$ active material for use as a cathode material in LIBs is routinely prepared through sol-gel, hydrothermal, or solid-state reactions [39, 40]. To date, to obtain the $LiFePO_4$/graphene composites, simple physical mixing, the hydrothermal method [41–43], the sol-gel method [44], the solid-state method [45], and the co-precipitation method [46] have been successfully employed. Ding et al. first reported a $LiFePO_4$/graphene composite with a specific capacity of 160 mAh g^{-1} as compared to 113 mAh g^{-1} for commercial $LiFePO_4$ [46]. Su et al. introduced graphene into $LiFePO_4$ as a planar conductive additive to produce a flexible graphene-based conductive network [47]. In these early studies, the authors reported that, even with a much lower fraction of graphene additive, the charge/discharge cycling performance of the $LiFePO_4$ nanocomposite electrode is better than those fabricated with commercial carbon-based additives. L. Wang et al. [42]

prepared a conducting $LiFePO_4$/graphene composite using a facile hydrothermal route followed by heat treatment. Morphological studies found that $LiFePO_4$ particles adhered to the surface of the graphene and/or became embedded in the GNS to form a 3D conducting network [42], which leads to good electronic conductivity and thereby better Li^+-ion diffusion. A $LiFePO_4$/graphene nanocomposite prepared in a ratio of 92:8 showed a discharge capacity of 160.3 mAh g^{-1} at 0.1 C and 81.5 mAh g^{-1} at 10 C due to an improved 3D conducting network–bridging GNS. To further improve the electronic conductivity of the composites, Y. Wang et al. prepared a $LiFePO_4$/graphene composite via a solid state route [48], and 3D porous $LiFePO_4$–graphene composites were synthesized by sol-gel [44], and mechanical mixing methods [49]. The composite prepared using solid state synthesis [48] reported a specific capacity of 161 mAh g^{-1} at 0.1 C and 50 mAh g^{-1} at 50 C, while the porous $LiFePO_4$/graphene composites showed the capacity values of 45–60 and 75–109 mAh g^{-1} without and with graphene, respectively, at a high rate of 10 C. It appears that the composite prepared using thee sol-gel method shows superior performance even at higher rates compared to those prepared by mechanical mixing and solid-state synthesis.

In an effort to increase the electronic conductivity of $LiFePO_4$, Dhindsa et al. [8] employed a sol-gel method to prepare a $LiFePO_4$/graphene nanocomposite. In this study, GO was mixed with $LiFePO_4$ precursors. The electron conductivity of the resulting composites was improved by six orders of magnitude compared to the pristine $LiFePO_4$ [8]. An electron microscope, scanning electron microscope (SEM), and transmission electron microscope (TEM) all confirmed that the $LiFePO_4$ particles uniformly covered by graphene sheets and forming a 3D conducting network were responsible for the high electronic conductivity. In addition, a high capacity of 160 mAh g^{-1} was recorded, which is close to the theoretical limit, and high stability in terms of the cathode was also observed. As for decreasing the distance of Li^+-ion diffusion, a 3D nanoporous micron-sized sphere is considered to be an optimal structure for $LiFePO_4$, capable of realizing high-power capability without compromising packing density and facilitating the fast and efficient transport of ion and charge [50].

In two different studies, a graphene/$LiFePO_4$ porous composite was prepared using a facile template-free sol-gel approach (Figure 7.1) [44] and facile precipitation method [51]. The obtained graphene/$LiFePO_4$ composite has a reversible capacity of 146 mAh g^{-1} at 17 mA g^{-1} after 100 cycles, which is 1.4 times higher than that of porous $LiFePO_4$ (104 mAh g^{-1}), demonstrating that the incorporated graphene greatly enhances the specific capacity throughout the cycle process. It was observed that a porous graphene/$LiFePO_4$ composite also exhibits a desirable tolerance to varied charge/discharge current densities. However, oversized holes fail to achieve the advantage of the high tap density of a porous structure, and the non-uniform distribution of graphene did not show significant improvement on electrochemical performance.

Based on this study, Ma et al. [51] successfully synthesized a composite using the as-prepared GO/Li_3PO_4 as a sacrificial template. The microspheres showed a uniform morphology that 2 μm $LiFePO_4$ microspheres are wrapped by GNS. The cathodes made from graphene/$LiFePO_4$ delivered excellent discharge capacities of

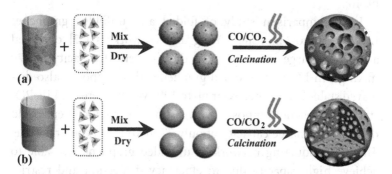

FIGURE 7.1 A schematic illustration showing the formation process of 3D porous networks for (a) $LiFePO_4$/graphene and (b) $LiFePO_4$. Reproduced from Scrosati et al. [44].

141, 130.9, and 101.8 mAh g^{-1} at current rates of 0.1, 1, and 10 C, respectively, due to the effective conducting network, formed by bridging GNS, and the porous structure, which could absorb the electrolyte to help Li$^+$-ion to diffusion. It shows excellent rate performance that holds 72% of the initial capacity at 10 C. The enhanced performance is attributed to the porous microsphere exhibiting a hierarchical structure assembled by nanoparticles, high conductivity and chemical stability of graphene network in composite and its porous structure in favor of Li$^+$-ion diffusion [51].

Ding et al. [46] also reported superior performance by nanostructured LiFePO$_4$/graphene composites prepared using a co-precipitation method in de-ionized water at room temperature. The GNS were used as scaffolds, with LiFePO$_4$ nanoparticles growing on the surface. The composites delivered an initial discharge capacity of 160 mAh g^{-1} at 0.2 C, and the capacity retained 110 mAh g^{-1} even at a high rate of 10 C with only 1.5 wt.% of graphene. The decrease in discharge capacity with a higher rate can be ascribed to the Li diffusion in the interface between LiFePO$_4$ and FePO$_4$. The rate performance and cycling stability of the LiFePO$_4$/graphene could be attributed to the fact that nanosized particle with surface area and improved conductivity through a superior graphene conductor and the reduced contact resistance [9, 46].

In a comparison study of folded and unfolded graphene/LiFePO$_4$ nanocomposites, Yang and co-workers [52] observed a 98% discharge capacity up to the theoretical capacity of 170 mAh g^{-1} in LIBs by unfolded graphene. The composite also displayed stable cycling behavior up to 100 cycles, whereas a LiFePO$_4$ stacked graphene composite with a similar carbon content could deliver a discharge capacity of only 77 mAh g^{-1} in the first cycle. The 3D conducting network of unfolded graphene was able to achieve high capacity due to efficiency dispersion and restriction of the LiFePO$_4$ particle size at the nanoscale. Unlike the folded graphene, which has a wrinkled structure, unfolded graphene makes it easier for each LiFePO$_4$ particle to attach itself to the conducting layer of graphene, which could greatly enhance electronic conductivity, thereby realizing the full potential of the active materials [52].

The specific capacity of the commercially available C-coated $LiFePO_4$ cathode is typically 120–160 mAh g^{-1}, which is lower than the theoretical value of 170 mAh g^{-1}. Recently, graphene-decorated C-coated $LiFePO_4$ with higher than 160 mAh g^{-1} has been reported [25, 53]. Hu et al. [25] reported a $LiFePO_4$ cathode with a specific capacity higher than the theoretical value using a C-coated $LiFePO_4$ surface modified with 2 wt.% of electro-chemically exfoliated graphene layers [25]. A specific capacity of 208 mAh g^{-1} was reported without causing unfavorable voltage polarization. The high conductivity of exfoliated graphene flakes wrapping around C-coated lithium iron phosphate assists electron migration during the charge/discharge cycles, leading to a 100% coulombic efficiency without fading at various C-rates. They also reported that the energy density is up to 686 Wh kg^{-1}, much higher than the typical 500 Wh kg^{-1} of $LiFePO_4$. The exceeded theoretical limit of the discharge capacity is attributed to the reversible reduction/oxidation reaction between the Li^+-ions of the electrolyte and the exfoliated graphene flakes, where the electrochemically exfoliated graphene flakes exhibit a capacity higher than 2,000 mAh g^{-1}. The highly conductive graphene flakes wrapping around C-coated $LiFePO_4$ also assist electron migration during the charge/discharge processes, diminishing the irreversible capacity at the first cycle and leading to ~100% coulombic efficiency [25]. However, even incorporating 3 wt.% graphene, the $LiFePO_4$ composite cathode shows only an initial discharge capacity of 164 mAh g^{-1}; this may be due to the clutching of Li^+-ions by the multilayer graphene films on $LiFePO_4$ nanosheets during the charge/discharge process [53].

The fabrication of a composite of graphene/$LiFePO_4$ has always proved cumbersome and time-consuming. The aim has always been to design an easy-to-use method that is scalable and low cost. Investigation of the hydrothermal method showed that it could be the best method in terms of the energy capacity, realizing 97% efficiency of the theoretical value [7]. In this method, the temperature was maintained at 180°C for 10 h, solution cooled at room temperature, precipitated, centrifuged, washed in deionized water three times, dried under vacuum for 4 h, and finally sintered at 600°C for 2 h under hydrogen/argon (5:95, v/v). Again, here, graphene was postulated to provide an easy pathway for faster electron transfer as well as Li^+-ion diffusion.

Another effective method for producing a graphene/$LiFePO_4$ composite is spray-drying. In this facile procedure, which was investigated by Zhou et al. [6], a viscous slurry prepared by the physical mixing of $LiFePO_4$ and reduced graphene oxide (rGO) suspension was spray dried at 200°C to form a solid $LiFePO_4$/graphene composite and finally annealed under argon to form the active $LiFePO_4$/graphene cathode materials. It was demonstrated that in the as-prepared composites, the $LiFePO_4$ primary nanoparticles embedded in micro-sized spherical secondary particles nanoparticles were wrapped homogeneously and loosely with rGO sheets, forming a 3D metal oxide–graphene network. Such a favorable structural motif facilitated electron migration throughout the secondary particles, while the presence of abundant voids between the $LiFePO_4$ nanoparticles and graphene sheets was beneficial for Li^+-ion diffusion. The as-prepared composite cathode showed a high specific capacity of 70 mAh g^{-1} at 60 C discharge rate and exhibited a capacity decay rate of <15% when cycled at a 10 C charging and 20 C discharging rate for 1000 cycles [6].

Sung et al. [54] demonstrated the synergistic combination of nanosizing with efficient conducting templates to afford facile Li^+-ion and electron transport for high-power applications. The authors reported on a size-constrained in-situ polymerization method to prepare core-shell C-coated $LiFePO_4$ nanoparticles hybridized with rGO as a cathode for high-power LIBs. Hydrophilic GO was employed as the starting template to preclude the spontaneous aggregation of hydrophobic graphene in aqueous solutions during the formation of a $LiFePO_4$/graphene composite cathode material via in-situ polymerization of pyrrole. Sucrose was mixed with PPy-coated $FePO_4$/rGO and annealed to get the carbon coating onto the active material. The authors ascribed the fabrication of true nanoscale C-coated $LiFePO_4$/rGO ($LiFePO_4$/C-rGO) hybrid cathode material to three factors: (i) in-situ polymerization of PPy for constrained nanoparticle synthesis of $LiFePO_4$, (ii) enhanced dispersion of conducting 2D networks endowed by the colloidal stability of GO, and (iii) intimate contact between the $LiFePO_4$ and rGO nanosheets. The authors demonstrated the importance of conducting template dispersion by contrasting $LiFePO_4$/C-rGO hybrids with $LiFePO_4$/C-rGO composites in which an agglomerated rGO solution was used as the starting template. The fabricated hybrid C-coated $LiFePO_4$/rGO cathodes showed superior rate capability and stable cycling performance at a low to high C-rate (0.05–60 C). The $LiFePO_4$/C-rGO cathode shows a discharge capacity close to the theoretical capacity of $LiFePO_4$ (170 mAh g^{-1}) at a slow rate of 0.05 C, and it delivers a capacity of 72 mAh g^{-1} at high C-rate of 60 C. The superior electrochemical performance of the resulting $LiFePO_4$-rGO hybrid was ascribed to the fast electron and ion transport through the nanosized active materials and 2D conducting scaffolds set up by the intimate contact formation between the active material and conducting networks, enhanced dispersion of conducting 2D networks, and size-constrained particle growth [54].

Bi et al. [55] used three types of strategies to prepare graphene, namely, chemical vapor deposition (CVD), a Wurtz-type reaction (a process using CCl_4, and substituting chlorine with conductive C=C layers), and chemical exfoliation [55]. The authors studied the electrochemical performances of $LiFePO_4$–graphene nanocomposite with the addition of 5% graphene as the cathode in LIBs and claimed reduced contact resistance between $LiFePO_4$ particles, which improved the overall electronic conductivity of the electrode and hence the electrochemical performance of the $LiFePO_4$. The study reported that the $LiFePO_4$–graphene nanocomposite prepared with graphene by CVD is more effective in terms of electronic conductivity, contact resistance, and electrochemical performance. The composite electrode is reported to have a specific capacity of 132 mAh g^{-1} and 80 mAh g^{-1} at 1 C and 20 C discharge rates, respectively. Solution-based approaches to making graphene/Li$_2$S composites have also been attempted. For instance, Wu et al. [56] showed that it is possible to prepare nano-sized Li$_2$S and graphene/Li$_2$S composites (Figure 7.2) that eliminate dangerous gases release.

Investigation of the LiMn$_2$O$_4$/graphene composite as cathode material have shown it to have higher capacity compared to pristine LiMn$_2$O$_4$. Kim et al. [57] showed that by varying the amount of graphene in a composite of LiMn$_2$O$_4$/graphene, the reversible capacity of the battery could be greatly improved. They further observed a reduced charge transfer resistance in the composite. In another study, using a simple vacuum filtration method, Ha and co-workers [5] were able to show that the oxygen content in rGO free-standing films for cathode applications plays a crucial role in the capacity of the battery. They showed that by increasing the oxygen functional groups, the capacity was also increased. Here, the Li-capturing mechanism was postulated to be caused by the interaction between Li$^+$-ions and surface oxygen functionalities and hence delivered excellent rate capability as well as higher surface Faradaic reaction.

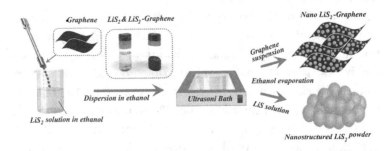

FIGURE 7.2 Schematic of the Li$_2$S and Li$_2$S-graphene composite synthesis process. Reproduced from Wu et al. [56].

Very recently, Tian et al. [58] and Wang et al. [59] have reported the use of a LiFePO$_4$/graphene aerogel in LIBs. Wang et al. [59] reported (010) on facet-orientated LiFePO$_4$ nanoplatelets wrapped in N-doped graphene aerogel [59], while the study by Tian et al. [58] focused on an undoped LiFePO$_4$/graphene aerogel. Wang et al. [59] also reported mesoporous C-coated LiFePO$_4$ nanocrystals co-modified with graphene and Mg^{2+} doping [59] as superior cathode materials for LIBs. In studies by Wang et al. [59], the composite cathode showed good cycling stability and rate capability. In the N-doped/Mn^{2+} doped composite cathode, the presence of graphene and the mesoporous structure offer rapid electron and Li$^+$-ions transport pathways, while the thin LiFePO$_4$ nanoplatelets/nanocrystals with a higher surface area enhance the active sites and shorten the Li$^+$-ion diffusion length. As a result, a capacity of 78 mAh g^{-1} at 100 C and good cycling stability over 1000 cycles are achieved. A capacity retention of ~90% is reported after 1000 cycles at 10 C [43, 59]. The graphene component quickly realizes a capacity response by creating a cushioning effect, thereby buffering the impact of Mg^{2+} doping even at a high charge/discharge rate, thereby resulting a good rate capability of 164 mAh g^{-1} at 0.2 C-rate, which is ~96% theoretical capacity; a capacity of 124 mAh g^{-1} at 20 C-rate; and excellent cycling stability [59].

7.3 FUNCTIONALIZED GRAPHENE AS CATHODE MATERIAL

Functionalized/doped graphene or reduced GNS have been studied as high-performance cathodes for future LIBs [60–62]. GO is reduced using ethylene glycole as both reducing agent and solvent [61], by thermal reduction [5, 60], and by N-doping [62]. The working electrode was prepared by mixing active material, conducting carbon, and poly(vinylidene fluoride) (PVdF) binder with a weight ratio of 80:10:10 using N-methylpyrrolidone (NMP) as the solvent. The coin cells were assembled in CR2032 using Li foil as the counter electrode and a solution of 1 M LiPF$_6$ in EC/DEC (1:1 v/v) as an electrolyte, and electrochemical performance was evaluated at the voltage range of 1.5–4.5 V (vs. Li/Li$^+$). At a current density of 50 mA g^{-1}, the cells delivered a charge/discharge capacity of about 220–250 mAh g^{-1} [60, 61]. By doping with nitrogen (oxygen 7.3% and nitrogen atomic 9.2%), the reversible capacity of GNS is improved up to 330 mAh g^{-1}, which is about 135 mAh g^{-1} higher than undoped GNS [62]. After 200 cycles, the functionalized GNS delivers a specific capacity of about 300 mAh g^{-1} [61], while the N-doped GNS delivers a specific capacity of 344 mAh g^{-1} [62] at a current density of 50 mA g^{-1}. After 300 cycles, the functionalized GNS delivers a specific capacity of 280 mAh g^{-1} at a current density of 50 mA g^{-1} [61], and the N-doped cathode delivered a high reversible capacity of 146 mAh g^{-1} after 1000 cycles at a current density of 1000 mA g^{-1} [62], suggesting that functionalization or doping marginally improves the cycling stability of graphene. The free-standing rGO electrode, having conducting additive and binder-free rGO with different oxygen contents, was assembled by a simple vacuum filtration process from aqueous rGO colloids prepared with the aid of cationic surfactants. The free-standing rGO electrode with C/O ratio of 15 was reported to have a specific discharge capacity of 125 mAh g^{-1} at the first cycle at 137 mA g^{-1} current rate, which corresponds to ~91% of the theoretical capacity of C/O = 15 cathodes [5]. It is observed that the capacity increased with the increase of the amount of oxygen functional groups, suggesting that the main Li-capturing mechanism of rGO cathodes is Li$^+$-ion interaction with surface oxygen functionalities. The hydroxyl groups (C-OH), as well as carbon–oxygen double bonds, have been identified as the lithiation active species [5, 60].

7.4 GRAPHENE QUANTUM DOTS AS CATHODE MATERIAL

Similar work using GNR with another material to form a composite for LIB application was also reported by Yang and co-workers [63]. Here, entrapping V$_2$O$_5$ nanocrystalline particles in GNR synthesized by the unzipping of carbon nanotubes (CNTs) was shown to improve electrical conductivity in LIBs when the composite was used as a cathode. Formation of a 3D conducting composite structure was thought to be responsible for the high cycle capacity, and the mass ratio of the two was controlled to result in a capacity of 278 mAh g^{-1} at 0.1 C with a maintenance of a high capacity of 165 mAh g^{-1} at 2 C.

7.5 GREEN ELECTRODES AS CATHODE MATERIAL

There has been a renewed interest in "green" cathode materials for LIBs. Initial investigations have shown these can represent alternative and cost-effective methods for the production of LIBs with comparable capacity. Polymers are the prime candidates for this purpose, two promising polymers being polyimide and poly(anthraquinonyl sulfide). In order to improve the high-rate performance of these polymers, in-situ polymerization in the presence of graphene has shown promising results where electronic conductivity was increased [64]. The poly(anthraquinonyl sulfide) composite cathode showed an ultrafast discharge/charge rate and was able to deliver in excess of 100 mAh g^{-1} in few seconds (16 s). A free-standing flexible cathode based on a self-assembled graphene poly(anthraquinonyl sulfide) composite aerogel with a specific capacity of 156 mAh g^{-1} at 0.1 C (1 C = 225 mAh g^{-1}) and an ultrahigh utilization (95%) of poly(anthraquinonyl sulfide) was demonstrated. This cathode exhibits outstanding cycle stability with ~85% capacity retention after 1000 cycles at 0.5 C and excellent rate performance with 102 mAh g^{-1} at 20 C in LIBs [65]. The comparison study of polyimide versus poly(anthraquinonyl sulfide) shows lower cathode utilization for a polyimide cathode. The cathode-based bare polyimide shows a polymer utilization ratio of 42% at 0.1 C, compared to the theoretical capacity of 367 mAh g^{-1}. By incorporating 6 wt.% graphene, the polyimide cathode showed a polymer utilization ratio of 49%, while poly(anthraquinonyl sulfide) shows a much higher polymer utilization ratio of about 88% [64].

Hydrophilic radical polymers have also been investigated as electrode materials for LIBs. In their study, Koshika et al. [66]

synthesized an electrode material using poly(2,2,6,6-tetrameth-ylpiperidinyloxy-4-yl vinylether). They reported a long life cycle and comparable high charging/discharging capability. The hydrophilic polymer was selected to prevent the low ignition risks associated with other materials.

Polyimides are also touted as suitable materials for electrochemical storage, especially in LIBs. It has been postulated that the aromatic imide group in polyimides can be reduced and oxidized electrochemically in a reversible manner, as proposed in Figure 7.3 [67]. As Figure 7.3 shows, this method can easily be described as enolization, showing a reverse process in the carbonyl group being promoted mainly by the conjugated structure through aliphatic [68] or aromaticity [69–71]. It is known that when polyimide is used as a cathode material in LIBs, its oxidation and reduction is accompanied by Li⁺-ion association and disassociation with oxygen. The Figure 7.3 further shows that it is possible to transfer four electrons in two steps, thereby enabling theoretical specific value above 300 mAh g^{-1} [67]. The composite organic cathode of polyimide nanostructures on graphene prepared by in situ polymerization showed good charge/discharge property cycling stability and rate capability [70, 71]. A high reversible capacity of 177–233 mAh g^{-1} at 0.1 C and an exceptionally high-rate cycling stability, that is, a high capacity of 109 mAh g^{-1} at a very high charge/discharge rate of 50 C with a capacity

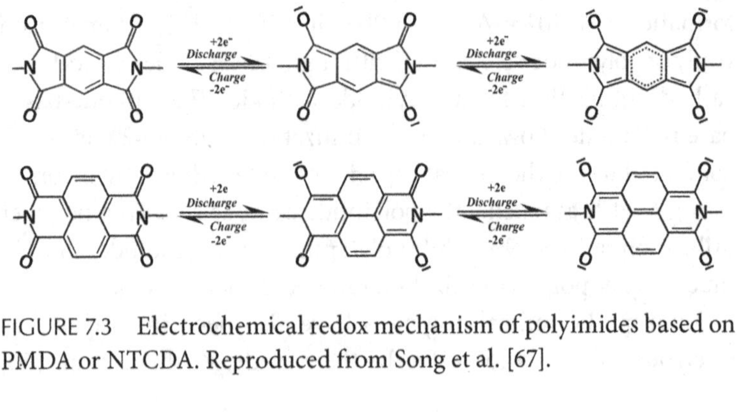

FIGURE 7.3 Electrochemical redox mechanism of polyimides based on PMDA or NTCDA. Reproduced from Song et al. [67].

retention of 80% after 1000 cycles, was reported. The polyimide cathode without graphene delivers only 164 or 185 mAh g^{-1} at a charge/discharge rate of 0.1 C. The significantly improved specific capacity, good cycling performance, rate capability, and low internal impedance of the cell are attributed to the synergetic effect of polyimide molecules on dispersed GNS. The in-situ polymerization generated intimate contact between polyimides and the electronically conductive graphene, resulting in conductive composites with highly reversible redox reactions and good structural stability. The study also demonstrates that the graphene-based composite also exhibited much better performance than composites based on MWNT and the conductive carbon black in terms of specific capacity and long-term cycling stability under the same charge/discharge rates [71].

Apart from conducting polymers, organic carbonyl or hydroxylated compounds with graphene have been reported in terms of a concept for future green and sustainable LIBs. A systematic study of decarboxylated GO and carbonylated/hydroxylated GO electrodes are reported to have a significantly enhanced electrochemical performance compared to GO electrodes. It is observed that carbonylated/hydroxylated GO exhibits much better electrochemical performance than conventional $LiCoO_2$ and $LiFePO_4$ cathodes; the higher specific capacity, good rate capability, and excellent cycling behavior is mainly due to the graphene-derived framework structures used for electron transport, together with an abundant carbonyl/hydroxyl functional group beneficial for Li^+-ion storage. Carbonylated/hydroxylated GO showed an average discharge capacity of 175 mAh g^{-1}, while GO and decarboxylated GO showed a capacity of 55 and 107 mAh g^{-1} respectively at a current density of 100 mA g^{-1}. The carbonylated/hydroxylated GO exhibits much higher capacity and cycling stability over 600 cycles, indicating that the manipulation of the oxygen functional groups on GO is an effective strategy for decreasing the irreversible lithiation process of GO-based electrodes while enabling high potential (vs. Li/Li^+) and significantly enhanced Li storage properties. The discharge capacity of carbonylated/hydroxylated GO at a current rate of 800 mA g^{-1} is as high as 84 mAh g^{-1}, which is close to two times that of a GO electrode (45 mAh g^{-1}) and over three times that of decarboxylated GO (24 mAh g^{-1}). Even after extended C-rate cycles, the high capacity of carbonylated/hydroxylated cathode at 100 mA g^{-1} can also be recovered to almost the initial value, suggesting that it has good reversibility. At the high current rates (400 and 800 mA g^{-1}), decarboxylated GO exhibits a lower capacity than GO, which is attributed to its long activating process. The excellent electrochemical performance of the carbonylated/hydroxylated GO is attributed to (i) its insolubility in the electrolyte, (ii) good conductivity for electron transport, and (iii) abundant hydroxyl and carbonyl groups for Li storage [72].

Rechargeable Li-S batteries are an economical alternative energy storage system due to the high theoretical capacity of sulfur. The poor electric/ionic conductivity of elemental sulfur or its Li compound (Li_2S/Li_2S_2), dissolution of intermediate long chain polysulfides into the electrolyte and their shuttle between cathode and anode leads to the inefficient utilization of active material, fast capacity degradation, and low coulombic efficiency. The precipitation of insoluble and insulating Li_2S/Li_2S_2 on the electrode surface due to the shuttle process causes loss of active material and renders the electrode surface electrochemically inactive. Among the different approaches, chemically bonding sulfur to the PAN backbone (S-PAN) is found to be a promising method for suppressing the shuttle effect. However, due to the poor electrical conductivity of PAN, the cycling stability and rate capability of S-PAN systems are still very modest. The incorporation of 3 wt.% rGO (44 wt.% sulfur) marginally improved the cycling stability and rate capability of the S-PAN composite cathode. The S-PAN/rGO-based LIB delivers an initial discharge capacity of 1827 mAh g^{-1} at 0.1 C-rate, which is much higher than the theoretical capacity of sulfur (1675 mAh g^{-1}). The extra capacity comes from the irreversible insertion of Li into the PAN backbone. In the second cycle, a reversible capacity of 1470 mAh g^{-1} was achieved with 90%

utilization of sulfur. The cell exhibits nearly 85% retention of the initial reversible capacity of 1467 mAh g^{-1} over 100 cycles at a current rate of 0.1 C and delivered a capacity of 1100 mAh g^{-1} after 200 cycles with an excellent coulombic efficiency, which implies a very minute amount of shuttle effect. At 2 C-rate, the cell delivers up to 828 mAh g^{-1} after 40 cycles. Compared to S-PAN/rGO, an S-PAN composite cathode decays rapidly, leading to a loss of ~25% capacity by the 80th cycle, and the cell is dead by the 100th cycle, which may due to the dendrite growth caused by the corrosion of the Li anode surface resulting from the polysulfide shuttle. The improved electrochemical performance of S-PAN/rGO is due to the good conductivity and high surface area imparted by rGO nanosheets, providing a robust electron transport framework. Also, the structural stability of the electrode imparted by rGO nanosheets enables the electrode to accommodate the large volume change in sulfur compounds during continuous expansion and contraction in the charge/discharge cycle, which prevents the cracking of the electrode. In addition, the functional groups on the rGO nanosheets wrapped around the polymer chain might help to prevent the dissolution of polysulfide by aiding PAN to maintain them within the electrode [73].

The search for polymer electrodes for LIBs has not been restricted to cathode electrodes alone; polymers that show redox [74] properties are prime candidates for this purpose. A composite blend of poly(2,5-dihydroxy-1,4-benzoquinone-3,6-methylene) and acetylene black showed a high specific capacity of 150 mAh g^{-1} when acetylene black loading was 40% [75]. It was further shown that the composite electrode exhibited excellent cyclability with a degradation loss of only 10% for the initial 100 cycles.

7.6 SUMMARY

LIBs continue to be the preferred choice in many applications due to their high energy capacity and ability to withstand many charge/discharge cycles with minimal energy loss. Optimizing the amount and properties of graphene and its derivatives as cathodes for LIBs will make it possible to power futuristic plug-in and hybrid electric vehicles due to their comparatively lower energy as well as power densities and inadequate durability along with high cost. This chapter summarized many innovative methods for synthesizing graphene-based cathode materials as well as various modifications that have been made to improve electrode performance. The large surface area, the nanostructured materials are capable of providing high Li^+-ion flux at the interface, abridged electronically conducting diffusion path for both ions and electrons, ample electrochemically active sites for Li^+-ion storage in addition to excellent accommodation for volume fluctuations associated with continuous charge/discharge cycles thereby boosting the structural integrity of the electrodes to prevent from the cracking. Even though the use of graphene in different cathodic materials eliminates some of the inherent problems associated with LIBs, graphene technology is much younger and yet to be established vis-a-vis cutting-edge battery technology.

REFERENCES

1. Song et al. *Mater. Sci. Eng. R Reports* 72 (2011) 203.
2. Fergus et al. *J. Power Sources* 195 (2010) 939.
3. Ritchie et al. *J. Power Sources* 162 (2006) 809.
4. Liu et al. *J. Nanomater.* 2013 (2013) 1.
5. Ha et al. *ACS Appl. Mater. Interfaces* 5 (2013) 12295.
6. Zhou et al. *J. Mater. Chem.* 21 (2011) 3353.
7. Lei et al. *Adv. Mater. Res.* 900 (2014) 242.
8. Dhindsa et al. *Solid State Ionics* 253 (2013) 94.
9. Chen et al. *J. Nanomater.* 2013 (2013) 1.
10. Qin et al. *J. Mater. Chem.* 22 (2012) 21144.
11. Gao et al. *ACS Appl. Mater. Interfaces* 6 (2014) 4154.
12. Wang et al. *J. Power Sources* 196 (2011) 7030.
13. Zhu et al. *Springerplus* 3 (2014) 1.
14. Shubha et al. *J. Power Sources* 267 (2014) 48.
15. Prasanth et al. *J. Power Sources* 202 (2012) 299.
16. Fisher et al. *Chem. Mater.* 20 (2008) 5907.
17. Andersson et al. *Solid State Ionics* 130 (2000) 41.

18. Prosini et al. *Solid State Ionics* 148 (2002) 45.
19. Choi et al. *J. Power Sources* 163 (2007) 1064.
20. Chung et al. *Nat. Mater.* 1 (2002) 123.
21. Chung et al. *Solid State Commun.* 131 (2004) 549.
22. Wagemaker et al. *Chem. Mater.* 20 (2008) 6313.
23. Herle et al. *Nat. Mater.* 3 (2004) 147.
24. Xu et al. *Mater. Lett.* 83 (2012) 27.
25. Hu et al. *Nat. Commun.* 4 (2013) 1687.
26. Kang et al. *New Carbon Mater.* 26 (2011) 161.
27. Oh et al. *Adv. Mater.* 22 (2010) 4842.
28. Zhao et al. *Electrochim. Acta* 55 (2010) 5899.
29. Pan et al. *Mater. Lett.* 65 (2011) 1131.
30. Koltypin et al. *J. Power Sources* 174 (2007) 1241.
31. Boyano et al. *J. Power Sources* 195 (2010) 5351.
32. Son et al. *J. Alloy. Compd.* 509 (2011) 1279.
33. Huang et al. *Chem. Mater.* 20 (2008) 7237.
34. Jugović et al. *Solid State Ionics* 179 (2008) 415.
35. Bai et al. *J. Alloy. Compd.* 508 (2010) 1.
36. Sun et al. *J. Phys. Chem. C* 114 (2010) 3297.
37. Kostecki et al. *Thin Solid Films* 396 (2001) 36.
38. Li et al. *J. Power Sources* 249 (2014) 311.
39. Xu et al. *Mater. Res. Bull.* 42 (2007) 883.
40. Xiang et al. *J. Power Sources* 195 (2010) 8331.
41. Zhang et al. *J. Power Sources* 210 (2012) 47.
42. Wang et al. *Solid State Ionics* 181 (2010) 1685.
43. Wang et al. *Energy Environ. Sci.* 8 (2015) 869.
44. Yang et al. *J. Power Sources* 208 (2012) 340.
45. Wang et al. *Mater. Lett.* 160 (2015) 210.
46. Ding et al. *Electrochem. Commun.* 12 (2010) 10.
47. Su et al. *J. Mater. Chem.* 20 (2010) 9644.
48. Wang et al. *Mater. Lett.* 71 (2012) 54.
49. Tang et al. *J. Power Sources* 203 (2012) 130.
50. Magasinski et al. *Nat. Mater.* 9 (2010) 353.
51. Ma et al. *Electrochem. Sci.* 8 (2013) 2842.
52. Yang et al. *Energy Environ. Sci.* 6 (2013) 1521.
53. Wang et al. *Carbon* 127 (2018) 149.
54. Ha et al. *Chem. A Eur. J.* 21 (2015) 2132.
55. Bi et al. *Electrochim. Acta* 88 (2013) 414.
56. Wu et al. *Part. Part. Syst. Charact.* 31 (2014) 639.
57. Kim et al. *Int. J. Electrochem. Sci.* 6 (2011) 5462.
58. Tian et al. *J. Power Sources* 340 (2017) 40.

59. Wang et al. *Nanoscale* 6 (2014) 986.
60. Xiong et al. *Electrochim. Acta* 174 (2015) 762.
61. Xiong et al. *Appl. Energy* 175 (2016) 512.
62. Xiong et al. *ACS Appl. Mater. Interfaces* 9 (2017) 10643.
63. Yang et al. *ACS Appl. Mater. Interfaces* 6 (2014) 9590.
64. Song et al. *Nano Lett.* 12 (2012) 2205.
65. Zhang et al. *ACS Appl. Mater. Interfaces* 9 (2017) 15549.
66. Koshika et al. *Green Chem. Lett. Rev.* 2 (2009) 169.
67. Song et al. *Angew. Chemie Int. Ed.* 49 (2010) 8444.
68. Armand et al. *Nat. Mater.* 8 (2009) 120.
69. Han et al. *Adv. Mater.* 19 (2007) 1616.
70. Ahmad et al. *RSC Adv.* 6 (2016) 33287.
71. Lyu et al. *ChemSusChem* 11 (2018) 763.
72. Ai et al. *Carbon* 74 (2014) 148.
73. Li et al. *J. Power Sources* 252 (2014) 107.
74. Oyaizu et al. *Adv. Mater.* 21 (2009) 2339.
75. Le et al. *J. Power Sources* 119 (2003) 316.

CHAPTER 8

Graphene/Polymer Nanocomposite Electrolytes for Lithium Ion Batteries

8.1 INTRODUCTION

If a high and reversible specific energy in lithium ion batteries (LIBs) is to be attained, it is essential that the repetitive deposition and stripping of Li remains highly reversible during the electrochemical process. Conventional Li^+-ion batteries containing organic liquid electrolytes are associated with the inherent disadvantages of leakage, problems with the solid electrolyte interface (SEI), electrical shorting, thermal runaway, and, in several cases, catastrophic fire hazards [1–3]. Since the cycling of Li metal is known to result in the deposition of Li dendrites that can decrease the cycle life of the cell and cause safety concerns, the development of suitable electrolytes that can suppress dendrite growth and improve plating morphology has been actively pursued [4].

There are many notable advantages to solid polymer electrolytes (SPEs), such as better thermomechanical and electrochemical properties as well as improved safety, especially for use in high-temperature applications; however, for room-temperature applications, SPEs are not suitable candidates due to their insufficient ionic conductivity at lower temperatures. Many methods, including the use of plasticizers [5–9], ceramic fillers [10–13], highly polar plastic crystals [14–17], and carbon nanostructures, have been employed to tackle the low room-temperature ionic conductivity of SPE without compromising its mechanical integrity; however, its practical applications in LIBs remain a nightmare.

The enhanced physical and electrochemical properties achieved are generally attributed to the large surface-to-volume ratio, robust mechanical strength, specific surface chemistry, and interfacial effects of nanofillers. The filler nanoparticles interact with polymer chains and electrolyte by the way of Lewis acid–base interactions [10–13], while plasticizers (e.g. EC/DMC) and plastic crystalline materials (e.g., N,N′-diethyl-3-methylpyrazolium bis[trifluoromethanesulfonyl imide] or butanedinitrile) considerably decrease the crystallinity of the composite material and act as a versatile Li$^+$-ion promoter due to their trans-gauche isomerism, involving the rotation of molecules about the central C-C bond [14, 18]. The trans isomers of plastic crystalline materials and ceramic fillers act as an impurity phase, which results in the enhancement of lattice defects and the lowering of the activation energy for ionic conduction, thereby affecting the ionic conductivity. Unlike ceramic fillers, which increase the conductivity of PEs by Lewis acid–base interaction between the filler and host matrix, plastic crystalline materials actively participate in the Li$^+$-ion conduction mechanism and are directly involved in an increase in ionic conductivity due to the presence of active Li$^+$-ion hopping centers [14, 19, 20]. In the process, the crystallinity of the host polymer is reduced and electrolyte dissociation also occurs.

The net result thus enhances the ionic conductivity. For example, compared to their microsized counterparts, nanosized alumina and silica with large surface densities are more efficient at improving the ionic conductivity of poly(ethylene oxide) (PEO) electrolyte by suppressing the crystallization of PEO [21–23].

It is also demonstrated that the incorporation of carbon nanotubes (CNTs) [24, 25] and 2D, single-atomic-thickness GO sheets, with their ultralarge surface area and excellent mechanical and electrical insulating properties, can be promising filler candidates for improving the ionic conductivity and mechanical properties of polymer electrolytes (PEs) [26–30]. Few studies have investigated the incorporation of GO/rGO sheets in a PEO host [31–33].

Graphene is a single-atom-thick, 2D sheet of sp^2 hybridized carbon, either graphene oxide (GO) or reduced graphene oxide (rGO). GO is made up of one or few layers of functionalized graphene derivatives containing oxygen-bearing functional groups on its edges and basal planes. GO has been utilized as functional filler material for preparing polymer nanocomposites (PNCs) for versatile applications, including PEs, owing to its unique chemical and physical properties. The usual properties of GO include high surface area disposability in organic solvents, as well as remarkable thermal/chemical stability and mechanical strength. The GO is electronically insulating, while rGO is conducting. GO sheets have a number of oxygenated functionalities, for example, epoxy hydroxyl and carboxyl, which can enhance ionic conductivity due to their good affinity to Li^+-ion through Lewis acid–base interaction, thereby further promoting the dissociation of Li salt into free Li^+-ions; its presence may also facilitate the establishment of continuous ion conducting channels within the PEs. It has been reported that the Li^+-ions can potentially find low-energy conducting paths along the interface of the GO and polymer matrix [24, 34]. In addition to facilitating high ionic conductivity, the presence of uniformly dispersed GO increases the mechanical stability of PEs due to the excellent mechanical property of GO sheets. Although rGO is known for its electrical conductivity, GO and functionalized GO exhibit insulating properties because sp^2 bonding networks are disrupted by the functionalization, which ensures the possible application of functionalized GO as an ion promoter-cum-filler material for PEs. In this respect, GO doped gel polymer electrolytes (GPEs) can be used for supercapacitors, solar cells, fuel cells, and Li^+-ion batteries [35, 36]. However, while only a few studies have reported on the use of GO as a filler for PEs for LIBs, surprisingly all of them are SPEs using PEO as the host matrix. In most of the studies, the SPE based on a PEO/GO nanocomposite is prepared using $LiClO_4$ [26, 31, 32, 37–41] as Li salt; however, a few studies have also been carried out using CF_3SO_3Li, LiTFSI and $LiNO_3$ [42] as Li salt.

8.2 PEO/GNS/LiClO$_4$ COMPOSITE SOLID POLYMER ELECTROLYTES

The unique properties of PEOs, such as their high dielectric constant and strong Li^+-ion solvating capability, makes them suitable for electrolyte application in LIBs. The use of GO sheets as nanofiller-cum-ion conducting promoter in a PEO matrix may promote the movement of the ethylene oxide (EO) segment at the EO/GO interface and influence the confined helix of the PEO-Li complex ion, which affects Li^+-ion mobility and thereby ionic conductivity. The swelling behavior of GO in polar solvent facilitates the intercalation of the host polymer into the lamellar GO sheets, thereby providing a uniform distribution of nanostructures in the PEO matrix. This, in turn, improves the Lewis acid–base interaction between the GO and PEO chains. PEO conducts ions mostly through its amorphous region via a hopping mechanism along the polymer segments, which is assisted by ether oxides. At room temperature, PEO homopolymer tends to crystallize and results in a much lower ionic conductivity than that required for batteries ($>10^{-4}$ S cm^{-1}). A few studies have reported on the use of a PEO/$LiClO_4$ nanocomposite SPE prepared by incorporating GO, rGO, and functionalized GO [26, 32, 37–41].

8.2.1 PEO/GNS/LiClO₄ Polymer Electrolytes for Room-Temperature Applications

In recent years, PEO-based PEs have received considerable attention as safe electrolytes for use in LIBs. The ionic conductivity and
thermal stability of PEO-based SPEs are significantly improved by
the incorporation of GNS as filler-cum-Li$^+$-ion-conduction promoter. Cheng et al. [37] analyzed the anisotropic ion transport
of solvent-casted PEO-LiClO$_4$ based SPE containing GO nanofiller and studied the enhancement in anisotropy ionic conductivity. It was found that by the slow evaporation of solvent, the
GO nanosheets become highly aligned parallel to the film surface;
this suppresses the crystallization of the PEO, which results in the
orientation of the polymer chain perpendicular to the film surface, that is, the PEO crystalline lamellae surfaces align parallel
to the film surfaces due to the synergistic effect of Li$^+$-ions and
GO. The reduction in the crystallization of polymer molecules
due to the incorporation of GO nanosheets serves as a template
for the crystal orientation and guides the ion transport, leading
to a highly anisotropic and conductive PNC electrolyte. The ionic
conductivity studies show anisotropy factors as high as ~70 for
the nanocomposite SPE, in which Li$^+$-ions may be enriched in the
vicinity of the GO surface due to the relatively strong interaction
between Li$^+$-ions and the polar functional groups on the GO surface. This surface Li$^+$-ion layer can further bind with PEO, leading
to only an amorphous PEO/Li$^+$-ion complex phase. This process
will significantly reduce the number of available nucleation sites
for PEO heterogeneous crystallization, leading to a decrease in the
crystallization kinetics and the formation of highly concentrated
Li$^+$-ion layers on the GO surface, which could result in a lower
ionic crosslinking of the PEO chain. In a nutshell, this study demonstrated that recrystallization can be tuned and controlled using
2D templates, furthering our understanding of the complex interactions that take place during ion transport at the fundamental
level, which can help guide the engineering of new and improved
PEO-incorporated batteries [37].

Gao et al. [32] and Yuan et al. [38] demonstrated the use of nanostructured GO as the filler in the PEO/LiClO$_4$ SPE. The study by Gao et al. [32] showed that the PEO matrix leads to the inflation of the free volume void and promotes the segmental motion of PEO at the polymer–nanofiller interface. A 70-fold increase in ionic conductivity is achieved at room temperature in the presence of a very small amount of GO loading (0.6 wt.%). GO sheets influence the confined helix of EO/Li$^+$-ions. The confined helix structure and the free volume site influence the Li$^+$-ion transfer; however, the free volume site is crucial in the conducting process. The enhancement in ionic conductivity is attributed to the fastest Li$^+$-ion conducting channels formed by the interconnected GO nanosheets at the polymer–filler interface. [32].

The studies conducted by Yuan et al. [38] showed an introduction of 1 wt.% GO into PEO/LiClO$_4$ SPE with EO/Li ratio of 16:1, the ionic conductivity being increased by two orders of magnitude (2×10^{-5} S cm^{-1}) and about 260% increment in tensile strength. In addition to the formation of interconnected conducting channels, the enhancement in ionic conductivity is attributed to (a) enhanced Li$^+$-ion mobility associated with improved segmental mobility of PEO chains due to the free-volume expansion and increased amorphous content of the PEO matrix, and (b) increased Li$^+$-ion concentration due to the filler-induced dissociation of the Li salt. Upon the addition of Li salt, the T_g of the electrolyte film is increased from −60°C to −20°C due to the salt effect resulting from the formation of complexes between PEO and LiClO$_4$; in contrast to this, the T_g of the PE decreases upon the incorporation of GO nanosheets. The T_g of SPE loaded with 5 wt.% GO is reported to have a T_g of −35°C, which is much lower than that of 1 wt.% GO-filled electrolyte. The reduction of the T_g of the electrolytes by the incorporation of GO sheets is due to the compact molecular packing and crystallization of the solid PEO, which is greatly disturbed by the presence of the GO nanosheets, resulting in the lowering of the T_g of the electrolyte. In general, a low T_g indicates higher free volume and amorphous content in the polymer matrix and higher molecular mobility and segmental motion of the polymer chains. Due to the intimate relation between ionic conductivity, T_g, free volume, amorphous content, and segmental mobility, a maximum ionic conductivity is expected for a nanocomposite electrolyte with higher GO loading. However, above 1 wt.% GO loading, the ionic conductivity declines, which suggests adverse filler effects at higher GO loading, counteracting the influence of reduced crystallinity and associated T_g, molecular mobility and segmental motion of the polymer. This adverse effect includes filler aggregation, diffusion tortuosity, and Li^+-ion trapping. A further improvement in ionic conductivity is achieved by incorporating 5% plasticizer; the plasticized PEO/GO electrolytes shows two order of magnitude improvement in ionic conductivity compared to unfilled plasticized PEO.

The feasibility study of PEO/GO/LiClO$_4$ SPE in coin cells using commercial electrodes (Graphitic anode and LiCoO$_2$ cathode) for room-temperature application is reported to have higher charge/discharge capacity and cycling stability for 70 cycles. The cell comprised of GO-filled electrolytes noticeably improved the performance of the battery, where the areal capacity reached 0.17 mAh cm^{-2}. The reasonable areal capacity of the Li$^+$-ion cell in the full-cell configuration demonstrates the practical application of GO-filled PE for thin film batteries. The mass density of the cathode material in the Li$^+$-ion cell was 0.012 g cm^{-2}. Since the Li$^+$-ion test cells contain commercial electrodes, which are more compactable with liquid than solid electrolytes, the mass capacity of the battery is relatively low, which would permit the development of high-performance paper-thin LIBs [38].

8.2.2 PEO/GNS/LiClO$_4$ Polymer Electrolytes for High-Temperature Applications

The thermal properties of the electrolytes are crucial for the safety of LIBs even at room-temperature applications, so it is inevitable for their use in LIBs for high-temperature applications. There are many reports on the enhancement of thermal properties of PEs by the incorporation of ceramic fillers [10–13], room temperature ionic liquids [8, 9, 43], CNTs [44, 45], other 2D materials such as nanoclay [46–48], and hexagonal boron nitride (hBN) [49], which have been reported for high-temperature applications up to 200°C. To the best of our knowledge, there are no reports on SPE-based PEO incorporated with GO for the high-temperature LIBs other than the studies by the authors of the book [26]. Figerez et al. [26] studied the Li$^+$-ion conductivity and electrochemical properties of PEO/GO/LiClO$_4$ SPE and evaluated its performance in LIBs at </= 80°C. The nanocomposite SPEs were prepared using an optimized amount of PEO/GO with varying LiClO$_4$ content using a mechanical mixing and evaporation casting method. Mechanical mixing is well known to be an effective way to prepare SPEs based on PEO-LiX, since it results in higher amorphous domains of PEO with small spherulitic morphology, leading to enhanced ionic conductivity, especially at lower temperatures [50]. The electrolyte films were prepared by mechanical mixing at a higher shear rate and an evaporation casting method for the fabrication of SPE based on PEO, resulting in thin free-standing non-tacky films with sufficient mechanical strength for easy handling [9].

Ionic conductivity studies on the AC impedance from room temperature to 80°C of PEO/GO incorporated with LiClO$_4$ SPEs showed that at room temperature, the impedance response of SPEs loaded with lower LiClO$_4$ content is in the form of a semicircle observed at medium to high frequency. For SPEs loaded with higher Li salt, a straight line inclines at about 40° with respect to the real axis at higher frequencies. This type of behavior is typically for SPEs sandwiched between two cohesive-blocking electrodes [51]. The impedance response is found to be typical of electrolytes where the bulk resistance is the major contribution to the total resistance and only a minor contribution from grain boundary resistance. The straight lines inclined toward the real axis representing the electrolyte electrode double-layer capacitance behavior were obtained for all the samples over the entire range of frequency evaluated from room temperature to 80°C.

The intercept on the real axis representing R_b of the electrolyte decreased to ~100 Ω with an increase in LiClO$_4$ [16] or LiTFSi [9] loading.

Yuan et al. [38] reported the room-temperature ionic conductivity of SPEs without GNS loaded with 13 wt.% of LiClO$_4$ to be 4×10^{-3} mS cm^{-1}. Compared to SPEs with 13 wt.% LiClO$_4$ without GO, the room-temperature ionic conductivity of SPEs incorporated with 2 wt.% of graphene is found to be 0.01 mS cm^{-1}, which is 150% higher than SPEs without GO [26, 39, 52], confirming the positive effect on the ionic conductivity of SPE on the incorporation of GO nanosheets. This sharp increment in the level of ionic conductivity is attributed to the ion transport properties of GO nanosheets. The 2D GO nanosheets have a number of oxygenated functionalities, including epoxy, hydroxyl, and carboxyl, which can exhibit good affinity for Li$^+$-ion transport and dissociate more LiClO$_4$ salt. When the GO is incorporated into the SPE, the GO nanosheets can interconnect to form a 3D network structure that can facilitate continuous Li$^+$-ion conducting channels within the SPE [53–56].

It has been reported that the temperature-dependent ionic conductivity (Arrhenius plot) of SPEs of PEO/GO/LiClO$_4$ in the range of room temperature to 80°C linearly increases with temperature. The enhancement in the ionic conductivity of SPE is very much pronounced at lower temperatures (<60°C). The conductivity sharply increases in the range from room temperature to 60°C due to the softening of PEO chains with a rise in temperature. The slope of the Arrhenius curve decreases above 60°C, which implies that the trending variation of ionic conductivity with temperature progressively changes from a well-defined two-region behavior for PEO/LiClO$_4$ SPE.

The change in the slope of the curve at nearly 60°C is typical of PEO-based electrolytes and has been attributed to the change in conduction mechanism of the SPE associated with the PEO crystalline-amorphous phase transition [57]. A similar observation of the enhanced ionic conductivity of PEO-LiTFSi has been reported, and the effect was attributed to the formation of salt containing an amorphous phase of PEO at low temperatures, leading to an overall reduction in the crystallinity of the system [58].

The ionic conductivity of the SPE is dependent on a number of charge carriers as per the equation $\sigma = \sum N_i R_i E_i$, where N_i, R_i, and E_i are the number of charge carriers, ionic charge, and ionic mobility, respectively. Higher swelling with temperature implies a larger number of free charge carriers, which translates to higher ionic conductivity. The Li^+-ions that are coordinated with the ether "O" atoms of the PEO segments could be freed partially or fully from that trap by coordinating with the anions (from the carboxylic group) present on GO nanosheets and hence could lead to a large number of charge carriers with improved charge migration. Thus, the incorporation of GO nanosheets is an effective way of enhancing the ionic conductivity of PEO-based SPEs to an approachable level, which has hitherto been not possible by other modifications [58]. A smooth and linear enhancement in ionic conductivity was observed with an increase in temperature from room temperature to 80°C. It was observed that within the temperature range, the Arrhenius plot is slightly curved, so that the activation energy for ionic conduction E_a can be obtained using the Vogel–Tamman–Fulcher (VTF) model, $\sigma = \sigma_0 T^{-1/2} \exp[-E_a/R(T-T_0)]$, instead of the simple Arrhenius model, $\sigma = \sigma_0 T^{-1/2} \exp(-E_a/RT)$, used for the treatment of the linear Arrhenius plots.

The practical application of SPEs in Li^+-ion cells (Li/SPE/LiFePO$_4$) at high temperature was reported with CR2016 coin cells. The cells were cycled at different C-rates between 2.5 and 4 V. The charge/discharge profiles from room temperature to 80°C showed that the SPE with a higher loading of $LiClO_4$ delivers higher specific capacity compared with lower Li salt concentration. It was also observed that the charge/discharge capacity and coulombic efficiency increases with temperature irrespective of Li salt concentration.

At higher temperatures, 100% coulombic efficiency and charge/discharge capacity is reached to the theoretical capacity of LiFePO$_4$

[31, 48, 49]. The difference in capacity of the SPE can be attributed to the difference in ionic conductivity of the SPE with varying Li-salt loading. The higher charge/discharge capacity delivered at higher temperatures is due to the increased conductivity and charge transfer number of SPE with temperature. Above 60°C, the PEO chains will undergo a melting process and start to lose their crystalline structure, leading to the facilitation of more soft Li^+-ion conduction channels.

The cycle performance and rate capability of $PEO/GO/LiClO_4$ showed stable charge/discharge properties even at higher C-rates at both room temperature and high temperature. As in PEO-based GPEs or SPEs, the charge/discharge capacity is well attained in the case of the SPE with a higher loading of Li salt. The result intern indicates that the SPE has good compatibility with both the electrodes (anode and cathode). The evaluation demonstrates the suitability of $PEO/LiClO_4$ SPE incorporated with rGO/GO nanosheets for room-temperature as well as high-temperature LIBs that can safely operate at elevated temperatures like 100°C.

The reports on rate capability studies showed good charge/discharge capacity and cycling performance even at higher C-rates, and a similar initial capacity is retained after the current density is changed back to a lower C-rate, which clearly demonstrates the high-capacity retention and rate capability of the cells. At a lower C-rate, the difference in capacity is more pronounced between the SPEs loaded with different $LiClO_4$ content. The higher ionic conductivity, lower interfacial resistance, better compatibility of the SPE is responsible for better cycling property and rate capability. At a fixed thickness, less conductive electrolyte results in an increase of the ohmic dopant, with the ion concentration gradient in the electrolyte leading to lower capacity [59, 60]. The lowering of capacity at a high C-rate is also attributed to the nature of the electrode. A $LiFePO_4$ cathode material is known to support the limitation of lower performance at a higher C-rate due to slow Li^+-ion diffusion at the solid two-phase boundary of $LiFePO_4/FePO_4$ [61].

8.2.3 PEO/GNS/LiClO$_4$ Polymer Electrolytes for Flexible Batteries

The thirst for developing flexible and rollable electronics, including ubiquitous touchscreens, wearable sensors, rolltops, and so on due to corporate competition in the electronics industry has fertilized research into flexible energy storage devices. Because of their superior attributes, LIBs are among the leading candidates to be transformed into paper-thin, flexible energy storage devices that can be embedded in textiles or directly attached to biological organs. The ultimate challenge in the evolutionary phase of flexible LIBs is to attain mechanical flexibility while maintaining the LIBs' high level of electrochemical performance. In this respect, among battery components, electrolytes represent the major challenge of meeting the requirements of burgeoning flexible electronic devices. Kammoun et al. [39] demonstrated the applicability of PEs exploiting the exceptional mechanical integrity and electrochemical properties of GO in flexible thin-film LIBs. In the study, 200 μm-thick PEs composed of 1 wt.% GO nanosheets embedded in a PEO matrix with LiClO$_4$ as the Li$^+$-ion species were prepared using the solvent casting method. The flexible LIBs were fabricated by stacking cathode, electrolyte, and anode onto two plastic sheets and laminating them. Before sealing the cell, a few drops of 1 M LiPF$_6$ in EC/DMC 1:1 v/v were added to the PE surface to enhance ionic conductivity and interfacial contact between the electrolyte and electrode. The capacity obtained for the PE was higher than that reported for a flexible battery comprised of a ceramic electrolyte [62], and it exhibited lower values compared to gel electrolyte-based flexible batteries [63, 64].

In addition to the higher specific capacity, the PEO/GO/LiClO$_4$ SPE provides enhanced safety and mechanical stability compared to gel electrolytes, which is primarily attributed to the presence of mechanically strong and thermally stable GO nanosheets and a negligible wt.% of a liquid plasticizer (5%–7%). Furthermore, the flexible LIB displays a high operating voltage of 4.9 V and an energy density of 4.8 mWh cm^{-3} at room temperature, which is within the range of values reported for thin-film LIBs (1–10 mWh cm^{-3}) [65]. The flexible LIB exhibits robust mechanical flexibility over 6000 bending cycles and good voltage retention as well as electrochemical performance in both flat and bent positions. Under the bent condition, the LIB with nanocomposite SPE delivered a higher average capacity (0.13 mAh cm^{-2}) than a flat battery. The bending induced through thickness compressive forces appeared to further enhance the contact between the SPE and the electrodes, which resulted in a decrease in the contact resistance between the electrodes and electrolytes of the battery. The cycling studies showed a coulombic efficiency of about 95%. The rate capability studies demonstrate good cycling stability, especially at a current density below 2 C, maintaining a high-capacity retention of more than 85%. The bending test also showed a high voltage retention of 93% after 6000 continuous bending cycles, which suggests that a relatively good adhesion exists between the nanocomposite SPE and the adjacent electrodes. The impedance analysis confirms that the presence of GO nanosheets reduces the internal and interfacial impedance of the battery. This demonstration of nanocomposite SPE with GO nanosheet-based flexible LIB paves the way for new forms of safer and more economically viable energy storage devices that would adapt to the stringent shape and space requirements of next-generation flexible electronic devices [39].

8.3 POLYMER GRAFTED GNS/LiClO$_4$ COMPOSITE POLYMER ELECTROLYTE

Many approaches have been taken to enhance the ionic conductivity and the electrochemical performance of PEO-based SPEs, including plasticization and the incorporation of ceramic particles and nanofiller such as carbon nanotubes (CNTs) and graphene. Among them, composite electrolytes incorporating graphene have shown superior properties. However, establishing a strong interfacial adhesion between the graphene surface and polymer matrix is difficult due to the poor dispensability of graphene into the polymer matrix; pristine GO usually establishes large interfacial interactions with polar polymers such as PEO. It is well known that chemical modifications or functionalization could improve the interfacial interaction between the polymer chains and carbon nanomaterials such as CNTs and graphene. The introduction of the proper functional group onto the graphene surface (grafting) could establish a strong interaction between polar polymers and GNS. The grafting of graphene is generally accomplished through a two-step procedure, the first step being oxidization and the second the attachment of amide or carbamate ester bonds to oxygen-containing groups of GO. Diwan et al. [66] studied the effect of GNS on the crystallinity and ionic conductivity of PEO/NH$_4$I electrolyte [66]. Based on this study, a few attempts have been made to functionalize GNS with small molecules, which could contribute in promoting ionic conduction; polyethylene glycol (PEG) is one of the most promising candidates for a plasticizer to be used in PEO-based electrolytes [67, 68]. The incorporation of plasticizers considerably increases the ionic conductivity and electrochemical properties of the electrolytes; however, it has a detrimental effect on the mechanical stability of SPEs. Based on these reports, Shim et al. [31] and Gomari et al. [41] made a couple of attempts to demonstrate that the incorporation of PEG-grafted graphene (PGO) could improve ionic conductivity and electrochemical properties while maintaining or even enhancing the mechanical strength of the SPE. Both studies used PEG-grafted GNS as the filler material and LiClO$_4$ as the Li salt for preparing the SPE. A study by Shim et al. [31] on composite SPE (EO/Li ratio = 0.07) uses an organic/inorganic hybrid branched-graft copolymer (BCP) based on poly(ethylene glycol) methyl ether methacrylate (PEGMA) and methacrylisobutyl POSS (MA-POSS) as the polymer matrix, while Gomari et al. [41] used PEO as the matrix material (EO/Li ratio is 8). Both SPEs were prepared using the solution casting process, with Shim et al. [31] using a filler loading of 0.2–10 wt.% and Gomari et al. [41] a 0.1–3 wt.% filler loading.

The ionic conductivity studies by Shim et al. [31] of the composite SPE based on BCP containing 0.2 wt.% of grafted GO $(2.1 \times 10^{-4}$ S cm^{-1} at 30°C) was found to be one order of magnitude higher than that of the SPE based on pristine BCP without filler $(1.1 \times 10^{-5}$ S cm^{-1} at 30°C). This phenomenon is attributed to the fact that a large amount of Li salt can be dissociated in the composite PE by Lewis acid–base interactions between the filler and the Li salt. It is reported that the SPE with grafted GO nanosheet content larger than 3 wt.% has lower ionic conductivity and where continuously decreased with further increase in filler loading 3 wt.%. The ionic conductivity of BCP/PGO was always higher than that of the BCP/GO. The ionic conductivities of both the SPEs with GO and PGO increased with filler loading up to a level of 0.2 wt.% for PGO and 0.5 wt.% for GO; however, when the GO and PGO content were lower than 0.2 wt.%, the ionic conductivity was found to be close to that of pristine BCP, and when the filler content was larger than 6 wt.%, the ionic conductivities were found to be even smaller than pristine BCP. It was also observed that the fractions of free ions of BCP/PGO are always larger than BCP/GO, demonstrating that the grafting of PEG chains onto the GO nanosheets is one of the most crucial factors in enhancing ionic conductivity. The PEG chains grafted onto the GO surface could solvate the Li$^+$-ions due to the presence of the ethylene oxide (EO) units, thereby increasing the amount of dissociated Li$^+$ free ions. In addition to the PEG chains, the carboxyl (COOH) and hydroxy (OH) groups of the GO can also interact with the Li salt by Lewis acid–base interactions, thereby further promoting the dissociation of the Li salt into Li$^+$ free ions. Because of this, the content of Li$^+$ free ions in BCP/PGO is always larger than that of SPE with/without GO. Another reason for the higher ionic conductivity of BCP/PGO than SPE with/without GO is attributed to the good dispersion state of PGO. The grafted GO showed much better dispersity in the host matrix than GO, because the PEG chains grafted onto GO nanosheets increase its compatibility with the polymer matrix. However, all the SPEs show similar electrochemical stability (about 5.3 V vs. Li/Li$^+$) at 60°C, irrespective of filler loading. This suggests that all the composite SPEs are electrochemically stable within the operation voltage range of 4 V class cathode materials such as V_2O_5 and $LiFePO_4$. The cell performance of all-solid-state Li$^+$-ion cells with pristine BCP and BCP/PGO with 2 wt.% of PGO at 60°C cycled at a C-rate of 0.1 C showed an initial discharge capacity of 287 mAh g^{-1} for BCP-PGO, which is close to the theoretical capacity of a V_2O_5 cathode. The higher charge/discharge property is attributed to the higher ionic conductivity of the BCP/PGO electrolyte, which contributes to the rapid Li$^+$-ion transport. The easy ionic transport alleviates the ohmic polarization loss of the cell, resulting in the larger specific capacity. Upon continuous cycling, about 30% of capacity fade was observed after ten cycles; however, no further significant capacity loss was observed. Similar initial capacity fade has been reported in all-solid-state LIBs [69]. The capacity fade in the all-solid-state battery may be due to an increase in charge transfer resistance, since the interfaces between the electrodes and SPE are relatively unstable owing to the solid nature of the electrolyte. In addition, for high-temperature batteries that are operated on or above 60°C, the binding capability of the poly(vinylidene fluoride) (PVdF) binder in the electrode is lost, leading to electrochemical performance deterioration in LIBs [70].

In studies by Gomari et al. [41], PEO/LiClO$_4$ matrix with molar ratio of eight between EO/Li shows an ionic conductivity of 1.3×10^{-6} S cm^{-1}, while on incorporating 0.1 wt.% GO, the ionic conductivity increases by ~500%, that is, 8.2×10^{-6} S cm^{-1}; and introducing 0.5 wt.% PGO gives an ionic conductivity of up to 2.5×10^{-5} S cm^{-1}, which is ~1800% enhancement in conductivity with respect to PEO/LiClO$_4$ and ~200% with respect to 0.1 wt.% GO-incorporated in PEO/LiClO$_4$ [41]. This is also 25% greater than the highest ionic conductivity achieved by any of the GO-incorporated PEO/LiClO$_4$ electrolyte [38].

8.4 PEO/GNS POLYMER ELECTROLYTES WITH DIFFERENT LITHIUM SALTS

The effect of GO in PEO along with three distinct Li salts—CF_3SO_3Li, LiTFSI, and $LiNO_3$—is demonstrated by Mohanta et al. [42], in which for all composite electrolytes, the ratio of EO:Li was maintained between 15:1. The electrolyte loaded with 0.5 wt.% graphene in the PEO matrix having CF_3SO_3Li showed an ionic conductivity of 9×10^{-6} S cm^{-1}, which is 175% greater than that of a pristine CF_3SO_3Li/PEO composite electrolyte (3.3×10^{-6} S cm^{-1}). The effect of GO on the enhancement of ionic conductivity of SPE is much more pronounced with LiTFSi and $LiNO_3$ than CF_3SO_3Li salt. The conductivity studies showed that when 0.3 wt.% GO is incorporated into a PEO complexed with a LiTFSi or $LiNO_3$ composite electrolyte, the ionic conductivity enhances from 2.5×10^{-6} to 6.8×10^{-4} S cm^{-1} and from 3×10^{-6} to 1.3×10^{-4} S cm^{-1}, respectively [42]. This enhancement in conductivity is attributed to a change in the PEO crystallinity, dipolar relaxation, and translational diffusion of the Li$^+$-ions.

8.5 SUMMARY

The electrochemical properties of GNS are explored as a functional filler for the development of PEs for room-temperature as well as high-temperature applications (>80°C), exploiting its exceptional thermal and electrochemical properties. Among various polymers, PEO is very attractive as the host matrix for making composite electrolytes with graphene. The presence of GNS not only improves the ionic conductivity by Lewis acid–base interaction but also actively participates in the Li$^+$-ion conduction mechanism by forming fast Li$^+$-ion conducting channels at the polymer–filler interface. The higher thermal and electrochemical stability offered by SPEs incorporating GNS make them a safe electrolyte for high-temperature LIBs. However, the possibilities of graphene in the electrolyte have been less explored. This chapter gives an overview of most of the available reports on polymer graphene composite electrolytes for room-temperature as well as high-temperature applications. The methodologies and approaches showed in the reports can be extended to produce a variety of PEs, achieving the required characteristics by choosing the appropriate combination of materials.

REFERENCES

1. Tarascon et al. *Nature* 414 (2001) 359.
2. Armand et al. *Nature* 451 (2008) 652.
3. Meyer et al. *Adv. Mater.* 10 (1998) 439.
4. Howlett et al. *Electrochem. Solid State Lett.* 7 (2004) 97.
5. Prasanth et al. *Mater. Res. Bull.* 45 (2010) 362.
6. Prasanth et al. *J. Power Source* 196 (2011) 6742.
7. Lim et al. *J. Nanosci. Nanotechnol.* 18 (2018) 6499.
8. Prasanth et al. *J. Power Source* 172 (2007) 863.
9. Prasanth et al. *Solid State Ionic* 178 (2007) 1235.
10. Prasanth et al. *J. Power Source* 184 (2008) 437.
11. Prasanth et al. *Electrochim Acta* 55 (2010) 1347.
12. Prasanth et al. *Electrochim. Acta* 54 (2008) 228.
13. Prasanth et al. *Mater. Res. Bull.* 48 (2013) 526.
14. Prasanth et al. *Electrochim. Acta* 125 (2014) 362.
15. Prasanth et al. *J. Power Source* 267 (2014) 48.
16. Prasanth et al. *J. Power Source* 245 (2014) 283.
17. Prasanth et al. *J. Power Source* 195 (2010) 6088.
18. Patel et al. *Elechem. Commun.* 10 (2008) 1912.
19. Abouimrane et al. *J. Electrochem. Soc.* 154 (2007) 1031.
20. Fan et al. *Adv. Fun. Mater.* 17 (2007) 2800.
21. Croce et al. *Philos. Mag. B Phys. Condens. Matter Stat. Mech. Electron. Opt. Magn. Prop.* 59 (1989) 161.
22. Kumar et al. *J. Power Sources* 52 (1994) 261.
23. Xu et al. *ACS Nano* 4 (2010) 5019.
24. Tang et al. *Nano Lett.* 12 (2012) 1152.
25. Ibrahim et al. *Solid State Commun.* 151 (2011) 1828.
26. Figerez, S.P. Graduate Thesis, Dept. of PSRT, CUSAT, April 2018.
27. Kotov et al. *Adv. Mater.* 8 (1996) 8637.
28. Cassagneau et al. *Adv. Mater.* 10 (1998) 877.
29. Kovtyukhova et al. *Chem. Mater.* 11 (1999) 771.
30. Paci et al. *J. Phys. Chem. C* 111 (2007) 18099.
31. Shim et al. *J. Mater. Chem. A* 2 (2014) 13873.
32. Gao et al. *J. Membr. Sci.* 470 (2014) 316.

33. Akhtar et al. *Nanoscale* 5 (2013) 5403.
34. Udomvech et al. *Chem. Phys. Lett.* 406 (2005) 161.
35. Shaheer et al. *Nanoscale* 5 (2013) 5403.
36. Yang et al. *Adv. Funct. Mater.* 23 (2013) 3353.
37. Cheng et al. *Macromolecules* 48 (2015) 4503.
38. Yuan et al. *RSC Adv.* 4 (2014) 59637.
39. Kammoun et al. *Nanoscale* 7 (2015) 17516.
40. Khan M.S., Shakoor A. *J. Chem.* 2015 (2015) 1.
41. Gomari et al. *Solid State Ionics* 303 (2017) 37.
42. Mohanta et al. *J. Appl. Polym. Sci.* 135 (2018) 22.
43. Prasanth et al. *Solid State Ionics* 262 (2014) 77.
44. Scrosati et al. *MRS Bull.* 25 (2000) 28.
45. Pasquier et al. *Solid State Ionics* 135 (2000) 249.
46. Prasanth et al. *Euro. Polym. J.* 49 (2013) 307.
47. Ajayan et al. *Sci. Rep.* 3 (2013) 2572.
48. Ajayan et al. *ACS Appl. Mater. Interfaces* 7 (2015) 2577.
49. Ajayan et al. *Adv. Energy Mater.* 6 (2016) 1600218.
50. Shin et al. *J. Power Sources* 107 (2002) 103.
51. Appetecchi et al. *J. Power Sources* 114 (2003) 105.
52. Das et al. *AIP Adv.* 5 (2015) 027125.
53. Cao et al. *J. Power Sources* 196 (2011) 8377.
54. Eda et al. *Adv. Mater.* 22 (2010) 2392.
55. Chen et al. *Chem. Soc. Rev.* 39 (2010) 3157.
56. Dreyer et al. *Chem. Soc. Rev.* 39 (2010) 228.
57. Joy et al. *Solid State Ionics* 178 (2007) 1235.
58. Shin et al. *J. Electrochem. Soc.* 152 (2005) 978.
59. Djian et al. *J. Power Sources* 187 (2009) 575.
60. Djian et al. *J. Power Sources* 172 (2007) 416.
61. Padhi et al. *J. Electrochem. Soc.* 144 (1997) 1188.
62. Koo et al. *Nano Lett.* 12 (2012) 4810.
63. Yang et al. *Proc. Natl. Acad. Sci. U. S. A.* 108 (2011) 13013.
64. Xu et al. *Nat. Commun.* 4 (2013) 1543.
65. Wu et al. *Nat. Commun.* 4 (2013) 2487.
66. Diwan et al. *Solid State Ionics* 217 (2012) 13.
67. Das et al. *Electrochim. Acta* 171 (2015) 59.
68. Kuila et al. *Mater. Sci. Eng. B* 137 (2007) 217
69. Prosini et al. *Electrochim. Acta* 46 (2001) 2623.
70. Yoon et al. *J. Power Sources* 215 (2012) 312.

Index

Printed in the United States
by Baker & Taylor Publisher Services

Printed in the United States
by Baker & Taylor Publisher Services